T0179144

PHARMACEUTICAL EXPERIMENTAL DESIGN AND INTERPRETATION

Second Edition

Pharmaceutical Experimental Design and Interpretation

Second Edition

N. Anthony Armstrong

University of Cardiff
UK

CRC Press
Taylor & Francis Group
Boca Raton London New York

CRC Press is an imprint of the
Taylor & Francis Group, an **informa** business

CRC Press
Taylor & Francis Group
6000 Broken Sound Parkway NW, Suite 300
Boca Raton, FL 33487-2742

First issued in paperback 2019

ISBN-13: 978-0-415-29901-5 (hbk)
ISBN-13: 978-0-367-39118-8 (pbk)

Library of Congress Cataloging-in-Publication Data available on application

Visit the Taylor & Francis Web site at
http://www.taylorandfrancis.com

and the CRC Press Web site at
http://www.crcpress.com

Author

Norman Anthony Armstrong graduated B.Pharm. and Ph.D. from London University. After some years in the pharmaceutical industry, Dr. Armstrong joined the Welsh School of Pharmacy, Cardiff University, U.K., where he became senior lecturer in pharmaceutical technology. He retired from that position in 2002.

Dr. Armstrong is a fellow of the Royal Pharmaceutical Society of Great Britain and is the author of over 150 scientific papers, reviews, and books.

Table of Contents

Chapter 1   Introduction to Experimental Design

1.1   The Experimental Process 1
1.2   Computers and Experimental Design 2
1.3   Overview of Experimental Design and Interpretation 4

Chapter 2   Comparison of Mean Values

2.1   Introduction 9
2.2   Comparison of Means when the Variance of the Whole
       Population is Known 10
2.3   Comparison of Two Means when the Variance of the Whole
       Population is Not Known 12
       2.3.1   Treatment of Outlying Data Points 15
2.4   Comparison of Means between More Than Two Groups
       of Data 18
       2.4.1   Analysis of Variance (ANOVA) 19
       2.4.2   The Least Significant Difference 21
       2.4.3   Two-Way Analysis of Variance 22

Chapter 3   Nonparametric Methods

3.1   Introduction 25
3.2   Nonparametric Tests for Paired Data 25
       3.2.1   The Sign Test 25
       3.2.2   The Wilcoxon Signed Rank Test 27
3.3   Nonparametric Tests for Unpaired Data 29
       3.3.1   The Wilcoxon Two-Sample Test 29

Chapter 4   Regression and Correlation

4.1   Introduction 33
4.2   Linear Regression 33
       4.2.1   The Number of Degrees of Freedom (Cell B11 in Table 4.4) 37
       4.2.2   The Coefficient of Determination (r^2) (Cell A10
               in Table 4.4) 38
       4.2.3   The Standard Errors of the Coefficients (Cells A9
               and B9 in Table 4.4) 40
       4.2.4   The F Value or Variance Ratio (Cell A11 in Table 4.4) 40
       4.2.5   The Two Regression Lines 41

4.3   Curve Fitting of Nonlinear Relationships 41
        4.3.1   The Power Series 42
        4.3.2   Quadratic Relationships 42
        4.3.3   Cubic Equations 43
        4.3.4   Transformations 44
4.4   Multiple Regression Analysis...44
        4.4.1   Correlation Coefficients..47
        4.4.2   Standard Error of the Coefficients and the Intercept......48
        4.4.3 F Value 48
4.5   Interaction between Independent Variables................................48
4.6   Stepwise Regression 49
4.7   Rank Correlation...50
4.8   Comments on the Correlation Coefficient 52

Chapter 5   Multivariate Methods

5.1   Introduction...55
5.2   Multivariate Distances 55
        5.2.1   Distance Matrices 55
5.3   Covariance Matrices 59
5.4   Correlation Matrices 62
5.5   Cluster Analysis...63
        5.5.1   Cartesian Plots 63
        5.5.2   Dendrograms 65
5.6   Discrimination Analysis 67
5.7   Principal Components Analysis 70
5.8   Factor Analysis 75

Chapter 6   Factorial Design of Experiments

6.1   Introduction...83
6.2   Two-Factor, Two-Level Factorial Designs 84
        6.2.1   Two-Factor, Two-Level Factorial Designs with
                Interaction between the Factors......................................86
6.3   Notation in Factorially Designed Experiments............................89
6.4   Factorial Designs with Three Factors and Two Levels 91
6.5   Factorial Design and Analysis of Variance.................................94
        6.5.1   Yates's Treatment..95
        6.5.2   Factorial Design and Linear Regression 98
6.6   Replication in Factorial Designs 100
6.7   The Sequence of Experiments..103
6.8   Factorial Designs with Three Levels..104
6.9   Three-Factor, Three-Level Factorial Designs............................110
        6.9.1   Mixed or Asymmetric Designs..114
6.10   Blocked Factorial Designs...115
6.11   Fractional Factorial Designs..118

6.12   Plackett–Burman Designs 121
6.13   Central Composite Designs 122
6.14   Box–Behnken Designs 126
6.15   Doehlert Designs 127
6.16   The Efficiency of Experimental Designs 129

Chapter 7   Response-Surface Methodology

7.1   Introduction 135
7.2   Constraints, Boundaries, and the Experimental Domain 136
7.3   Response Surfaces Generated from First-Order Models 137
7.4   Response Surfaces Generated by Models of a Higher Order ... 143
7.5   Response-Surface Methodology with Three or More Factors ... 150

Chapter 8   Model-Dependent Optimization

8.1   Introduction 157
8.2   Model-Dependent Optimization 158
        8.2.1   Extension of the Design Space 161
8.3   Optimization by Combining Contour Plots 163
8.4   Location of the Optimum of Multiple Responses by
        the Desirability Function 165
8.5   Optimization Using Pareto-Optimality 168

Chapter 9   Sequential Methods and Model-Independent Optimization

9.1   Introduction 173
9.2   Sequential Analysis 173
        9.2.1   Wald Diagrams 173
9.3   Model-Independent Optimization 177
        9.3.1   Optimization by Simplex Search 177
9.4   Comparison of Model-Independent and
        Model-Dependent Methods 184

Chapter 10   Experimental Designs for Mixtures

10.1   Introduction 189
10.2   Three-Component Systems and Ternary Diagrams 190
10.3   Mixtures with More Than Three Components 193
10.4   Response-Surface Methodology in Experiments with Mixtures ... 195
        10.4.1   Rectilinear Relationships between Composition
                    and Response 195
        10.4.2   Derivation of Contour Plots from Rectilinear Models ... 197
        10.4.3   Higher-Order Relationships between Composition
                    and Response 198
        10.4.4   Contour Plots Derived from Higher-Order Equations ... 200

10.5   The Optimization of Mixtures...202
10.6   Pareto-Optimality and Mixtures...203
10.7   Process Variables in Mixture Experiments 205

Chapter 11   Artificial Neural Networks and Experimental Design

11.1   Introduction...209
          11.1.1   Pharmaceutical Applications of ANNs 212

Appendix 1   Statistical Tables

A1.1   The Cumulative Normal Distribution (Gaussian Distribution)................219
A1.2   Student's *t* Distribution...219
A1.3   Analysis of Variance...221

Appendix 2   Matrices

A2.1   Introduction...223
A2.2   Addition and Subtraction of Matrices..225
A2.3   Multiplication of Matrices...226
          A2.3.1   Multiplying a Matrix by a Constant..226
          A2.3.2   Multiplication of One Matrix by Another..................................226
          A2.3.3   Multiplication by a Unit Matrix 227
          A2.3.4   Multiplication by a Null Matrix 228
          A2.3.5   Transposition of Matrices 228
          A2.3.6   Inversion of Matrices 229
A2.4   Determinants 229

Index...233

Dedication

for

Kenneth Charles James, 1926–1997

The first two editions of this book were written in collaboration with Dr. Kenneth Charles James, reader in pharmaceutics at the Welsh School of Pharmacy, Cardiff University. Sadly, just as the second edition was being completed, Ken's health deteriorated and he died shortly after its publication.

This edition is therefore dedicated to the memory of Ken James, mentor, colleague, and friend.

1 Introduction to Experimental Design

1.1 THE EXPERIMENTAL PROCESS

Experimentation is expensive in terms of time, work force, and resources. It is therefore reasonable to ask whether experimentation can be made more efficient, thereby reducing expenditure of time and money.

Scientific principles of experimental design have been available for some time now. Much of the work originated with Sir Ronald Fisher and Professor Frank Yates, who worked together at Rothamsted Agricultural Research, U.K.[1] The principles that they and others devised have found application in many areas, but it is surprising how little these principles have been used in pharmaceutical systems. The reasons for this neglect are a matter of speculation, but there is no doubt that principles of experimental design do have a widespread applicability to the solution of pharmaceutical problems.

Experimentation may be defined as the investigation of a defined area with a firm objective, using appropriate tools and drawing conclusions that are justified by the experimental data so obtained. Most experiments consist of measuring the effect that one or more factors have on the outcome of the experiment. The factors are the independent variables, and the outcome is the response or dependent variable.

The overall experimental process may be divided into the following stages:

1. Statement of the problem. What is the experiment supposed to achieve? What is its objective?
2. Choice of factors to be investigated, and the levels of those factors that are to be used.
3. Selection of a suitable response. This may be defined in Stage 1, statement of the problem. If so, then we must be sure that the measurement of the chosen response contributes to achieving the objective. The proposed methods of measuring the response and their accuracy must also be considered at this stage.
4. Choice of the experimental design. This is often a balance between cost and statistical validity. The more an experiment is replicated, the greater the reliability of the results. However, replication increases cost, and the experimenter must therefore consider what is an acceptable degree of uncertainty. This in turn is governed by the number of replicates that can be afforded. Inextricably linked with this stage is selection of the method to be used to analyze data.

5. Performance of the experiment: the data collection process.
6. Data analysis.
7. Drawing conclusions.

The steps in the process may be illustrated using a simple example that is developed further in Chapter 4. Gebre-Mariam et al.[2] investigated the relationship between the composition of mixtures of glycerol and water and the viscosity of those mixtures, as part of a study of diffusion through gels.

Thus, the objective (Stage 1) was to establish the dependence of the viscosity of glycerol–water mixtures on their composition. The factor to be investigated (Stage 2) was composition of the mixture up to a maximum of about 40% w/w glycerol. The response (Stage 3) was the viscosity of the liquids, measured by an appropriately accurate method, in this case a U-tube viscometer. Because only one factor was to be investigated, any other factor that might influence the response had to be eliminated or kept constant. Temperature was an obvious example in this case.

At the outset, it was not known whether the relationship would be rectilinear or curvilinear. Furthermore, results were to be fitted to a model equation, and for both these reasons, an adequate number of data points had to be obtained. Five concentrations of glycerol were selected, covering the desired range (Stage 4). This was expected to be the minimum number that would enable a valid regression analysis to be performed. Many data points could have been used, thereby improving the reliability of any relationship, but of course this would have involved additional work.

The experiments were then carried out (Stage 5), the data was subjected to regression analysis (Stage 6), and the relationship between composition and viscosity was established (Stage 7).

Thus, the experimental design and the method to be used to analyze the data are selected before the experiment is carried out. Conclusions that can be drawn from the data depend, to a large extent, on the manner in which the data were collected. Oftentimes, the objective of the experiment is imperfectly defined, the experiment is then carried out, and only after these are methods of data analysis considered. It is then discovered that the experimental design is deficient and has provided insufficient or inappropriate data for the most effective form of analysis to be carried out. Thus, the term experimental design must include not only the proposed experimental methodology, but also the methods whereby the data from the experiments is to be analyzed. The importance of considering both parts of this definition together cannot be overemphasized.

1.2   COMPUTERS AND EXPERIMENTAL DESIGN

A point that must be considered at this stage is the availability of computing facilities such as mainframes, personal computers (PCs), and even a pocket calculator. The advantages of the computer are obvious. The chore of repetitive calculation has been removed as well as an undeniable disincentive to use statistical methods. However, using a computer can cause two related problems. The first is absolute reliance on the computer — if the computer says so, it must be so. The second is the assumption that the computer can take unreliable data or data from a badly designed experiment and somehow transform them into a result which can be relied upon. The computer jargon GIGO — garbage in, garbage out — is just as appropriate to problems of experimental design as to other areas in which computers are used.

It is undeniable that access to a computer is invaluable. Many readers will have access to a mainframe computer equipped with comprehensive statistical packages including SPSS® (McGraw-Hill, New York, NY, USA), SAS® (SAS Institute, Cary, NC, USA), and MINITAB® (Minitab, State College, PA, USA). Bohidar[3] has described the application of SAS to problems of pharmaceutical formulation.

MINITAB contains many features that are relevant to experimental design. In addition to useful statistical techniques, it includes programs for determinant analysis and principal component analysis (Chapter 5). The commands FFDESIGN and PBDESIGN generate fractional factorial designs and Plackett–Burman designs respectively for a specified number of experimental factors (Chapter 6). Randomization of the order in which the experiments are to be performed can also be carried out. The command FFACTORIAL analyzes data from experiments based on these designs, and facilities for drawing contour plots from the data are also available (Chapters 7 and 8). Details are given in Ryan and Joiner.[4]

However, a desktop computer will suffice for many of the calculations described in this book, because many statistical packages for PCs are now commercially available. Spreadsheet packages such as Lotus 1-2-3® (Lotus Development Corporation, Cambridge, MA, USA) and Excel® (Microsoft Corporation, Redmond, WA, USA) are of great value for these calculations.[5] The latter is used extensively in this book.

Several software packages specifically intended for experimental design and optimization purposes are also available. One example is the RS/Discover® suite of programs from BBN Software Products Corporation (Cambridge, MA, USA). The menu-driven program in this package prompts the user to specify the independent variables, together with their units, the ranges of values for the variables, and the required degree of precision and to indicate whether the value of a given variable can be easily altered. The program then produces a worksheet that gives the design of the experiment (full factorial, central composite, etc.) and the values of the independent variables for each experiment. The experiments are usually given in random order, except in those cases where a particular experimental variable cannot be easily altered in value. In such cases, the experiments are grouped so that the time taken to alter that variable is minimized. After the experiments are carried out, the responses are added to the worksheet. Data can then be analyzed and fitted to models and contour plots, and response surfaces can be produced. Applications of this package have been reported by McGurk et al.[6]

The Design-Ease® and Design-Expert® packages offered by Stat-Ease (Minneapolis, MN, USA) provide facilities for the design and analysis of factorial experiments. The programs generate worksheets of experiments in random order or in blocks for experiments involving process variables or mixtures and, from the results, can produce a statistical analysis and three-dimensional response surface and contour graphs.

Similar programs include ECHIP® (Expert on a Chip, Hockessin, DE, USA), which has been reviewed by Dobberstein et al.,[7] CHEOPS® (Chemical Operations by Simplex, Elsevier Scientific Software, Amsterdam, The Netherlands), Statgraphics Plus® (Statgraphics, Rockville, MD, USA), and CODEX® (Chemometrical Optimisation and Design for Experimenters, AP Scientific Services, Stockholm, Sweden).

1.3   OVERVIEW OF EXPERIMENTAL DESIGN AND INTERPRETATION

This is not a textbook on statistics. However, some statistical knowledge is essential if the full power of techniques in experimental design is to be appreciated. Neither is this a compendium of methods of experimental design. Rather, it discusses methods that are of value in the design of experiments and in the interpretation of results obtained from them.

The literature in this area is considerable, and for readers wishing to develop their knowledge of a particular technique, references to further reading are given at the end of each chapter. Moreover, statistical textbooks and some general texts on experimental design are given at the end of this chapter.

Many experiments consist in acquiring groups of data points, each group having been subjected to a different treatment, and methods for evaluating data from such experiments are included in Chapter 2. Essentially, these methods are based on establishing whether the mean values of the various groups differ significantly. When there are only two groups of data, Student's t-test is usually applied, but for three or more groups, analysis of variance is the method of choice. The latter also forms the basis of many of the methods of experimental design described in later chapters.

For Student's t-test and analysis of variance to be applicable, the data should, strictly speaking, be normally distributed about the mean and must have true numerical values. Such tests cannot be applied to adjectival information or when data have been assigned to numbered but arbitrarily designated categories. In such cases, nonparametric methods come into their own. These methods do not depend for their validity on a normal or Gaussian distribution, and "adjectival" data can be assessed using them. However, such methods depend on the presence of an adequate number of data points to facilitate comparison, and hence the degree of replication in the experiment must be appropriate if such methods are to be used. Nonparametric methods involve either paired data, where each subject acts as its own control, or unpaired data. Both are discussed in Chapter 3.

Having obtained raw data from the experiment, one has to decide on how best to use them. The decision may be simple; for example, all that is required is a mean value and standard deviation or the plot of one value against another, which gives a perfect straight line. Usually, more is required, in which case the statistical method that is most appropriate to the problem must be chosen.

An obvious example involves a series of pairs of results where it is required to know whether they are related, and if so how. A simple example could be the variation of the weights of a collection of laboratory animals with their heights. A plot of height (h) against weight (w) drawn on a graph paper may not give a definite answer, because the points could be such that it is not clear whether or not the results are scattered around a straight line. The probability that the results are so related is given by regression analysis, together with the value of the line in predicting unknown results. Alternatively, the relationship may be curved but fits a quadratic equation.

If the results are not related, a third property, for example, age (A), may make an important contribution. It is not possible to plot a graph in this situation, although one could construct a three-dimensional model.

It is not possible to visually represent equations with more than three variables, but such higher relationships can be expressed by an equation. Thus, for example, if the variation of animals' weights (w) with height, age (A), and waist circumference (c) is examined, a relationship of the form shown in (1.1) can be devised:

$$w = b_0 + b_1 h + b_2 A + b_3 c \tag{1.1}$$

in which b_0, b_1, b_2, and b_3 are constants and can be derived by regression analysis. A minimum of four sets of data (because there are four variables) would be required to derive such an equation, and a perfect relationship would result. For a reliable relationship, a minimum of five sets of data for each unknown, giving a minimum of 20 sets of results, are necessary.

Other relationships can be detected, either by trial and error or by suspected relationships, derived theoretically or found for similar systems in the literature; for example, logarithmic (1.2), ternary (1.3), or square root (1.4). Some examples are given in the book, and methods for calculating them and evaluating their reliability are described.

$$y = b_0 + b_1 \log x \tag{1.2}$$

$$y = b_0 + b_1 + b_2 x^2 + b_3 x^3 \tag{1.3}$$

$$y = b_0 + b_1 x^{1/2} \tag{1.4}$$

Regression analysis looks for relationships between a dependent variable and one or more independent variables. This method of analysis is called a univariate method. Multivariate methods look for relationships between several variables, considering them collectively. These data are often presented in the form of a matrix, an example of which follows:

$$\begin{bmatrix} a_1 & a_2 & a_3 & a_4 \\ b_1 & b_2 & b_3 & b_4 \\ c_1 & c_2 & c_3 & c_4 \\ d_1 & d_2 & d_3 & d_4 \end{bmatrix} \tag{1.5}$$

Each column represents a property of the materials under examination. For example, 1 could represent tablet weight, 2 disintegration time, 3 crushing strength, and 4 moisture content. Each row represents a combination of the properties of one example, in this case the properties of a different tablet formulation. To work with these, one must have a knowledge of matrices and their manipulation, which differs from basic algebraic methods. The basic matrix algebra necessary to understand this section is given in Appendix 2, followed by examples of their use.

When a series of results is presented, the individual results can frequently be arranged into unrelated groups, within which the results are related. This is called cluster analysis. Alternatively, the validity of preconceived classifications can be examined by discrimination analysis.

Relationships within sets of results can often be detected and used to simplify data. Thus, the number of rows shown in (1.5) could possibly be reduced to three or even less by principal components analysis and the columns reduced in a similar manner by factor analysis. Cluster, discrimination, principal components, and factor analysis are all described in Chapter 5.

Experimental programs can, if not efficiently designed, consume much time, materials, and labor, and hence, it is essential that programs be designed in the most cost-effective manner. In Chapter 6, the principles of factorial design are discussed. Factorial design, when allied to statistical techniques such as analysis of variance, is a powerful tool for gaining the maximum amount of information from a limited number of experiments.

Factorial design involves the variation of two or more experimental variables or factors in a planned manner, and the factors are investigated at two or more levels. The technique establishes the relative order of importance of the factors and can also indicate whether factors interact and whether such interactions are significant.

Even so, full factorial designs involving several factors at three or even more levels can demand considerable resources. Therefore, methods by which the number of experiments can be reduced in factorial designs are also explored. The potential hazards of using such limited designs are also discussed.

Having determined which factors and interactions make a significant contribution to the response, one can use the same experiments to predict the response for combinations of factors that have not been studied experimentally. The prediction is carried out by deriving a mathematical model relating the factors to the response. The construction of the model equation and establishing its validity draw heavily on correlation and regression techniques described in Chapter 4.

Once the model is established, it can be used to construct contour plots. These plots are diagrams of the value of the response in terms of the values of the experimental variables. The model can also be used to derive the response surface. This is usually a three-dimensional diagram, with the response plotted on the vertical axis and two factors forming the horizontal axes. Such diagrams are invaluable in visualizing relationships between independent and dependent variables and also in assessing the robustness of the response. Both are described in Chapter 7.

Many pharmaceutical formulations and processes lend themselves to optimization procedures, whereby the best possible result is sought, given a series of limits or constraints. Thus, the best possible solution is not necessarily a maximum (or minimum) value, but is rather a compromise, taking many factors into account. There are two principal methods of optimization. One is model-dependent optimization, in which a group of experiments is carried out and the results are then fitted to an equation (the model). Such techniques are discussed in Chapter 8.

Model-dependent methods require that a series of experiments should be carried out and the results assessed only when the whole series has been completed. Methods by which the results of only a few experiments govern the conditions of further experiments are sequential or model independent, and the results are examined continuously as they become available. No attempt is made to express results in a model equation. Such methods are described in Chapter 9, which also includes a comparison between model-dependent and model-independent techniques.

Many pharmaceutical formulations involve mixtures of several ingredients, the total mass or volume of which is fixed. The composition of a fixed-volume injection or the contents of a hard-shell capsule are good examples. Here, if the proportion of one ingredient is changed, then the proportion of at least one of the others must also change. Such mixtures are amenable to the principles of experimental design, the applications of which are described in Chapter 10.

In the final chapter, the use of artificial neural networks in pharmaceutical experimental design is considered (Chapter 11). Artificial neural networks are machines that learn from experience, in a similar manner to the brain. Their underlying function is to identify patterns, that is, to recognize the relationship between input data and the corresponding response. These relationships are then applied in a predictive manner.

Each chapter is illustrated by a number of worked examples. Their selection has sometimes caused problems. Inevitably the author has tended to select examples which he has found of value, and which are therefore in fields in which he is personally interested. However he accepts that there are many other areas of pharmaceutical science that could have been explored. Therefore, many of the chapters end with a bibliography that indicates those areas where a particular technique has been used, and the reader is referred to the original articles.

The appendices of the book contain material to which reference may be required, but which would be intrusive if it was contained in the main body itself. Tabulated statistical data (e.g., values of Student's t-test, F-test, and correlation coefficients at given significance levels) has been reduced to a minimum and only includes material that is needed in the worked examples used in the book. Complete tables are readily available elsewhere.

Useful Statistical Texts

Bolton, S. and Bon, C., *Pharmaceutical Statistics: Practical and Clinical Applications*, 4th ed., Marcel Dekker, New York, 2004.

Clarke, G. M. and Cooke, D. A., *A Basic Course in Statistics*, 4th ed., Arnold, London, 1998.

Jones, D. S., *Pharmaceutical Statistics*, Pharmaceutical Press, London, 2002.

Useful General Texts on Experimental Design

Anderson, V. L. and McLean, R. A., *Design of Experiments: A Realistic Approach*, Marcel Dekker, New York, 1974.

Box, G. E. P., Hunter, W. G., and Hunter, J. S., *Statistics for Experimenters: Introduction to Design, Data Analysis and Model Building*, Wiley, New York, 1978.

Cornell, J. A., *Experiments with Mixtures*, 3rd ed., Wiley, New York, 2002.

Fisher, R. A. and Yates, F., *The Design of Experiments*, 8th ed., Oliver & Boyd, Edinburgh, 1966.

Hicks, C. R. and Turner, K. V., *Fundamental Concepts in the Design of Experiments*, 5th ed., Oxford University Press, Oxford, 1999.

Montgomery, D. C., *Design and Analysis of Experiments*, 5th ed., Wiley, New York, 2001.

Strange, R. S., Introduction to experiment design for chemists, *J. Chem. Educ.*, 67, 113, 1990.

REFERENCES

1. Fisher, R. A., *The Design of Experiments*, Oliver & Boyd, London, 1926.
2. Gebre-Mariam, T. et al., The use of electron spin resonance to measure microviscosity, *J. Pharm. Pharmacol.*, 43, 510, 1991.
3. Bohidar, N. R., Pharmaceutical formulation optimization using SAS, *Drug Dev. Ind. Pharm.*, 17, 421, 1991.
4. Ryan, B. F. and Joiner, B. L., *Minitab Handbook*, 4th ed., Duxbury Press, Pacific Grove, 2001.
5. Dranchuk, J., *Excel for Windows Spreadsheet Databases*, Wiley, New York, 1994.
6. McGurk, J. G., Storey, R., and Lendrem, D. W., Computer-aided process optimisation, *J. Pharm. Pharmacol.*, 41, 128P, 1989.
7. Dobberstein, R. H. et al., Computer-assisted experimental design in pharmaceutical formulation, *Pharm. Technol.*, 3, 84, 1994.

2 Comparison of Mean Values

2.1 INTRODUCTION

A common feature of many experimental programs is to obtain groups of data under two or more sets of experimental conditions. The question then arises: Has the change in experimental conditions affected the data? The question may be rephrased to a more precise form: Do the means of each group differ significantly or are all groups really taken from the same population, the change in experimental conditions having had no significant effect? A variety of experimental techniques exist to answer this question. Hence, it is all too easy to select an inappropriate technique, with misleading results.

For selecting the correct procedure, further questions must be asked:

1. Are the data truly numerical? Some data are purely nominal, in that they are given a name, for example, male or female, black or white. Such data, especially if they are to be processed by a computer, are often given a numerical value, for example, male = 0, female = 1, but these are labels, not actual numbers. Data can also be ordinal, in that they are ranked. For example, five children can be ranked in order of increasing height, with the value of 1 assigned to the shortest child and 5 to the tallest. These are not truly numerical values, in that the series does not represent a scale with equal intervals. Thus, there is no suggestion that the difference in height between numbers 1 and 2 is the same as that between 2 and 3. If, however, the actual heights of the children had been used, then these are truly numerical data and can be used in the tests described below.
2. Are there more than two sets of data?
3. Are the data normally distributed?
4. Are there many data points in each group (more than 30)?
5. If there are only two sets of data, do these sets represent the total population or do they represent samples drawn from a larger population? Do we know the variance of the whole population? Examples of the former could be sets of examination results, when the performance of every candidate is known. Also, in a long-running industrial process, where many batches have been made under identical conditions, the pooled variance of all the batches will be very close to or even equal to the variance of the total population or universe.
6. Are the data paired or unpaired?

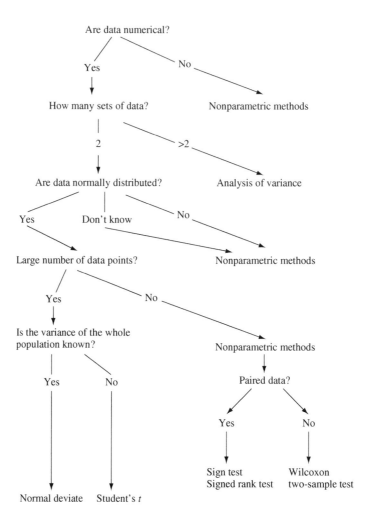

FIGURE 2.1 Chart to help select the correct statistical test for comparison of the means of groups of data.

Figure 2.1 shows Questions 1 to 6 in a diagrammatic form.

The available procedures can best be illustrated by examples which, though apparently straightforward, will serve as media through which several aspects of experimental design can be explored.

2.2 COMPARISON OF MEANS WHEN THE VARIANCE OF THE WHOLE POPULATION IS KNOWN

Twenty university students are taught a given subject in two groups of ten (Groups A and B), each group having its own tutor. At the end of the course, all 20 students take the same examination, the results of which are shown in Table 2.1.

TABLE 2.1
Marks Obtained by Two Groups of Ten Students (%)

	Group A	Group B
	70	66
	60	56
	59	55
	56	53
	56	48
	54	45
	52	45
	51	44
	44	42
	44	38
n	10	10
Mean	54.6	49.2
Variance	53.4	61.8
Standard deviation	7.3	7.9

The means differ by over 5% on marks of about 50%, which seems quite large. On the other hand, the values of the standard deviations show that there is considerable scatter around each mean. The university is concerned by the difference in mean marks between the two groups and wishes to assess whether this difference is statistically significant.

Figure 2.1 shows that use of the normal deviate is an appropriate test, because the data relate to the whole of the population and not just to samples. The procedure is to use the normal deviate to construct confidence intervals for the means. The confidence interval for Group A is given by (2.1):

$$\text{confidence interval} = x_{\text{mA}} \pm \left(\frac{Z_P \sigma}{\sqrt{n_A}} \right) \tag{2.1}$$

where
x_{mA} = mean of Group A
σ = standard deviation of Group A
n_A = number of observations in Group A
P = required level of probability
Z_P = normal deviate corresponding to the $(P+1)/2$ percentile of the cumulative standard normal distribution.

Thus, a key point to be decided is the required level of probability, as this governs the value of Z_P, and this decision must be taken before the calculation can be made. In most physicochemical experiments, a significance level of 0.05 is selected, which means that there is a 1 in 20 chance of the wrong inference being made.

Table A1.1 in Appendix 1 summarizes a selection of values of the standard normal variable. The value to choose is that corresponding to $(P+1)/2$, which in this case is 0.975. Therefore, $Z = 1.96$.

Hence, the confidence interval for the mean of Group A is

$$54.6 \pm \frac{1.96 \times 7.3}{\sqrt{10}} = 54.6 \pm 4.5 = 50.1 \text{ to } 59.1$$

The mean of Group B falls outside this range; therefore, it can be concluded that, at this level of significance, there is a difference between the means. It may be, however, that the university foresees serious consequences if a significant difference between the performances of the two groups is established. It therefore decides to choose a significance level of 0.01, so that there is now only a 1 in 100 chance of an incorrect inference being made. The chosen value of the standard normal deviate now corresponds to $(P+1)/2 = 0.995$. Therefore, the tabulated value of Z is now 2.58. Substituting this into (2.1) gives the confidence interval for the mean of Group A as 48.6 to 60.6. The mean of Group B lies within this range, and it could be claimed that there is no significant difference between the means. Thus, whether a significant difference exists depends on the level of significance that is chosen. This, in turn, is selected with the consequences of drawing the wrong conclusion firmly in mind.

2.3 COMPARISON OF TWO MEANS WHEN THE VARIANCE OF THE WHOLE POPULATION IS NOT KNOWN

In the previous example, every member of the population (all 20 students) was tested. In many cases, however, this is not feasible. The total population may be too high for it all to be tested or the testing may be destructive. In such cases, the variance must be estimated from data obtained from samples.

As an example, consider the following situation. Hard-shell capsules are filled with a mixture of active ingredients and diluents (Formulation A). A new formulation is devised (Formulation B) which, it is believed, will alter the disintegration times of the capsules. The objective of the experiment is therefore to establish whether a significant difference exists between the mean disintegration times of the two formulations. The capsules are subjected to the disintegration test of the European Pharmacopoeia, and the results are given in Table 2.2.

Figure 2.1 indicates that the appropriate test in this case is Student's t-test. There are two formulae that can be used for calculating t. The first is (2.2):

$$t = \frac{x_{mA} - x_{mB}}{\sqrt{s_p^2 \left(1/n_A + 1/n_B\right)}} \tag{2.2}$$

where

x_{mA} and x_{mB} = means from Formulations A and B, respectively

n_A and n_B = the number of data points in each group

TABLE 2.2
Disintegration Time (Minutes) of Hard-Shell Capsules
Containing Two Formulations, A and B

	Formulation A	Formulation B
	11.1	9.2
	10.3	10.3
	13.0	11.2
	14.3	11.3
	11.2	10.5
	14.7	9.5
n	6	6
Mean	12.43	10.33
Variance	3.36	0.74
Standard deviation	1.83	0.86

$S_p^2 =$ the pooled variance, which is, in turn, given by (2.3)

$$s_p^2 = \frac{(n_A - 1)s_A^2 + (n_B - 1)s_B^2}{(n_A + n_B - 2)} \qquad (2.3)$$

where
S_A^2 and $S_B^2 =$ the variances of the data from Formulations A and B, respectively.

Alternatively, t can be calculated from (2.4)

$$t = \frac{x_{mA} - x_{mB}}{\sqrt{\left(s_A^2 / n_A + s_B^2 / n_B\right)}} \qquad (2.4)$$

Equation (2.2) is used when the variances of the two sets of data do not differ considerably. A ratio between the variances of less than 3 is a good rule of thumb. If the variances differ by more than this, (2.4) is used instead. Use of (2.4) gives a more conservative estimate of significance than (2.2), even when both samples have similar variances. For the data shown in Table 2.2, the ratio of the variances is 4.8 (3.4:0.7); therefore, (2.4) is used to calculate t.

Statistical tests such as Student's t involve comparison of a value of t calculated from the data with a tabulated value. If the calculated value exceeds the tabulated value, then a significant difference between the means of the two groups has been detected. Tabulated values of t are shown in Table A1.2 in Appendix 1. Before the correct tabulated value can be selected, two items of information are required, which are in turn dependent on the design of the experiment. The first is the required level of significance, that is, the required value of P

in the top row of Table A1.2 in Appendix 1. The usual value of P for physico-chemical experiments is 0.05.

The second decision to be taken before the experiment can be carried out is the number of replicate determinations that will be made. The higher the number of replicates, the higher will be the cost of the experiment. On the other hand, an increase in replication increases the number of degrees of freedom shown in the first column of Table A1.2 in Appendix 1. Table A1.2 shows that the tabulated value of t falls as the number of replicates increases. Hence, even if the mean and variance of the measurements remain the same, a significant difference is more likely to be detected with an increase in the number of replicates.

The calculated value of t is also altered by changing the number of replicates. Equation (2.2) and Equation (2.4) show that if the number of degrees of freedom is increased, the calculated value of t will rise, and a significant difference between the means is again more likely to be detected.

After these decisions are taken, the experiment can be carried out. In accordance with the disintegration test of the European Pharmacopoeia, six measurements are carried out on each formulation, and the results are shown in Table 2.2. Because the ratio of the variances is greater than 3, t is calculated using (2.4), giving a calculated value of 2.540 for t. There are 10 degrees of freedom (n_A-1+n_B-1), and hence, the tabulated value at $P=0.05$ for a two-tail test is 2.228. Because the calculated value of t is greater than this, a significant difference exists between the mean disintegration times of the two formulations at this level of probability.

It is worthwhile exploring the conclusions that might have been drawn if the experiment had been designed differently concerning the chosen probability level and the number of replicates, assuming that, irrespective of the number of measure-ments, the means and standard deviations remain unchanged from the values given in Table 2.2. Table 2.3 shows the effect of increasing the value of n on both the calculated and tabulated values of t. The conclusion that a significant difference exists between the means is confirmed at $P=0.05$. If a different level of significance is chosen, for example, $P=0.01$, then with only six determinations per formulation

---

TABLE 2.3

Changes in the Calculated and Tabulated Values of t with Increased Replication, Assuming that the Means and Standard Deviations of the Data in Table 2.2 Remain Unchanged

Measurements in Each Group ($n_A = n_B$)	Calculated t	Degrees of Freedom	Two-Tail Test[a] $P=0.05$	Two-Tail Test[a] $P=0.01$	One-Tail Test[a] $P=0.05$	One-Tail Test[a] $P=0.01$
6	2.540	10	2.228	3.169	1.812	2.764
12	3.593	22	2.074	2.819	1.717	2.508
18	4.400	34	2.042	2.750	1.697	2.457
24	5.081	46	2.021	2.704	1.684	2.423

[a] Data are tabulated t

it is not possible to support the view that the mean disintegration time has changed significantly. However, it does become possible if the disintegration times of 12 capsules of each formulation are measured.

This example has investigated whether or not a significant difference exists between the mean disintegration times of the two formulations. This is a two-tail test, as the value of $m_A - m_B$ can be either positive or negative. However, a claim might have been made that Formulation B disintegrated more quickly, that is, m_B would be less than m_A, and so $m_A - m_B$ would always be expected to be positive. This is a one-tail test. The calculated values of t remain the same, but the tabulated values must take into account that this is now a one-tail test. The critical value of t at a level of significance of $P=0.05$ and 10 degrees of freedom is now 1.812. Hence, it can be concluded that Formulation B gives capsules with a significantly shorter disintegration time.

2.3.1 TREATMENT OF OUTLYING DATA POINTS

Measurements are inherently variable, and occasionally a result may be obtained which is very different from that which was expected. This is termed an outlying result. The way in which outlying results are to be treated is another aspect that should be considered before the experiment is carried out, rather than after the results have been obtained. Consider the data in Table 2.4. These are identical to those in Table 2.1, except that the last result in Group B (38) is replaced by zero. This has a marked effect on the mean and variance of Group B's results and hence on the conclusions that can be drawn from the experiment as a whole. Different conclusions would be drawn if that result were to be left out of the calculation.

TABLE 2.4
Marks Obtained by Two Groups of Students (%)

	Group A		Group B		
	10 Students	8 Students	10 Students	8 Students	9 Students
	70	—	66	—	66
	60	60	56	56	56
	59	59	55	55	55
	56	56	53	53	53
	56	56	48	48	48
	54	54	45	45	45
	52	52	45	45	45
	51	51	44	44	44
	44	44	42	42	42
	44	—	0	—	—
n	10	8	10	8	9
Mean	54.6	54.0[a]	45.4	48.5[a]	50.4
Variance	53.4	22.8	276.8	25.8	53.1
Standard deviation	7.3	4.8	16.6	5.1	7.3

[a] Trimmed mean

There are several possible sources of error in obtaining experimental data; for example, human, instrumentation, and calculation errors may occur. However, as a rule, data should not be rejected unless there is a very good reason for doing so. In this case, the question must be asked: why did that particular student score zero? An explanation for that outlying result must be sought. If a zero score was obtained because he or she did not attend the examination, then omission of that mark is justified. If, however, the student took the test in the same way as the others, then there is no justification for leaving that mark out of the calculation.

Techniques are available for identifying and dealing with outlying data points. One such technique is to calculate what is known as a trimmed mean, which disregards a proportion of the highest and lowest values in the data set. Table 2.4 gives trimmed information for the two groups in which the highest and lowest data points in each group are omitted from the calculation. The problem that arises from this approach is that perfectly valid data (the highest value in Group B and the extreme values in Group A) are no longer part of the calculation. Reducing the number of data points [the denominator in (2.1)] has the unavoidable consequence of widening the confidence interval and hence increasing the difficulty of establishing a significant difference between the groups.

Several more objective tests to identify outlying data points are available. One such test eliminates data points that are more than four standard deviations from the mean, because it is extremely unlikely to find such a value ($P < 0.00005$) so far from the mean. On the basis of this, none of the results from Group B would be identified as an outlier, because four standard deviations (4×16.6) from the mean (45.4) would give a range of -21.0 to 111.8. The presence of an outlying result inevitably increases the standard deviation. In addition, the probability level quoted above is derived on the assumption that the data are normally distributed. This is unlikely if potential outliers are present, unless there are many data points.

Another method of identifying outlying results is to construct what is known as a box plot (Figure 2.2).

Figure 2.2 is constructed as follows. The lower and upper quartiles (Q_1 and Q_3) of the data are identified. These are 44 and 55, respectively, and from these, the

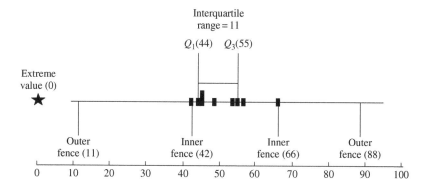

FIGURE 2.2 Box plot derived from data in Table 2.4, Group B.

interquartile range $(Q_3 - Q_1)$ is calculated. This value of 11 constitutes the central "box." From either side of the box, lines are extended as far as the last observation within 1.5 times the interquartile range. These extremities are called the inner fences of the plot. The lines are then extended further to 3 times the interquartile range, giving the outer fences. Any value lying between the inner and outer fences is considered to be a potential outlier, and any value outside the outer fence constitutes an extreme value. The box plot is a method of identifying outlying data points. It is not suggested that all extreme values should be discarded, but rather they should be scrutinized to ensure that they do form part of that data set. There is one extreme value (0) in Group B and none in Group A.

A third test for identifying outlying data points is Hampel's rule. In this process, the data are normalized by subtracting the median from each data point, and the absolute values of these deviations are noted, that is, the sign of the deviation is disregarded. Use of the median avoids any assumption that the data are normally distributed. The median of these absolute deviations is calculated and multiplied by 1.483 to give the median of absolute deviations (MAD). The absolute value of each deviation is divided by the MAD to give the absolute normalized deviation. Any result greater than 3.5 is considered an outlier. The process applied to the data from Group B is shown in Table 2.5. The median of the data points is 46.5, and the median of the absolute deviations is 5.5. Hence, the MAD is $5.5 \times 1.483 = 8.16$. Only the final data point (0) has an absolute normalized deviation greater than 3.5.

The use of tests to identify outlying data points is discussed in the 28th edition of *United States Pharmacopoeia*.[1]

TABLE 2.5
Identification of Outlying Data Points Using Hampel's Rule

Group B (10 Students)	Deviation from Median	Absolute Deviation	Absolute Normalized Deviation
66	19.5	19.5	2.42
56	9.5	9.5	1.18
55	8.5	8.5	1.06
53	6.5	6.5	0.81
48	1.5	1.5	0.19
45	−1.5	1.5	0.19
45	−1.5	1.5	0.19
44	−2.5	2.5	0.31
42	−4.5	4.5	0.56
0	−46.5	46.5	5.78
Median 46.5		5.5	
MAD		8.16	

Note: MAD = median of absolute deviation.

2.4 COMPARISON OF MEANS BETWEEN MORE THAN TWO GROUPS OF DATA

The examples discussed so far involve the comparison of the means of only two groups of data. However, there may be more than two groups. Consider the following example.

Tablets are made using three different formulations: A, B, and C. A sample of ten tablets is selected from each batch and the crushing strength of each tablet measured. The data are given in Table 2.6. Do the mean crushing strengths differ significantly?

A possible way forward would be to carry out multiple t-tests, that is, compare Batch A with Batch B, Batch B with Batch C, and Batch C with Batch A. The results of this are

$$\text{Batch A and Batch B: } t = 1.51$$
$$\text{Batch A and Batch C: } t = 3.07$$
$$\text{Batch B and Batch C: } t = 1.80$$

Thus, the mean of Batch C is significantly different from that of Batch A at a probability level of $P = 0.05$, the tabulated value of t with 18 degrees of freedom being 2.10.

There is a serious flaw in this approach: a probability level of 0.05 means that in 95% of cases, the statement associated with that level will be correct. In 5% of cases it will be wrong. Now three probability statements have been made, and if there is a 5% chance of each being wrong, then there is a 15% chance of one of the

---

TABLE 2.6
The Crushing Strengths of Tablets (kg) from Batches A, B, and C

	Batch A	Batch B	Batch C
	5.2	5.5	3.8
	5.9	4.5	4.8
	6.0	6.6	5.1
	4.4	4.2	4.2
	7.0	5.6	3.3
	5.4	4.5	3.5
	4.4	4.4	4.0
	5.6	4.8	1.7
	5.6	5.3	5.9
	5.1	3.8	4.8
n	10	10	10
Total	54.6	49.2	41.1
Mean	5.46	4.92	4.11
Grand total		144.9	
Variance	0.59	0.69	1.34
Standard deviation	0.77	0.83	1.16

three being wrong. Furthermore, there is no way of knowing which result is incorrect. Thus, as the number of groups of data increases, there is a rapidly diminishing chance of a correct overall assessment being made using the t-test. The proper way to proceed in these circumstances is to use analysis of variance (ANOVA).

2.4.1 ANALYSIS OF VARIANCE (ANOVA)

ANOVA is an extremely powerful statistical tool, permitting the comparison of the means of several populations. It assumes that a random sample has been taken from each population, that each population has a normal distribution, and that all the populations have the same variance. In practice, the last two requirements are not essential if sample sizes are approximately equal. The question that ANOVA seeks to answer is, are there significant differences among the means of the groups?

Obviously, within each group of data there will be scatter, and there will also be scatter between the groups. The variation within a group is an unexplained variation arising from random differences between the subjects and sources of variation that either are unknown or are being ignored. The problem is to answer the question: is the between-group variation significantly greater than the within-group variation?

The ANOVA procedure is best approached as a series of numbered steps, using as an example the data given in Table 2.6:

1. Calculate the total and the mean of every column.
2. Calculate the grand total. (The results of these first two steps are in Table 2.6.)
3. Calculate the (grand total)^2/(number of observations)
   $=(144.9)^2/30=699.87$.
   This term is used several times in this calculation. It is often called the correction term and denoted by the letter C.
4. Calculate the sum of (every result)^2
   $=(5.2)^2+(5.9)^2+\cdots+(4.8)^2=732.71$.
5. Subtract C from the result of Step 4
   $=732.71-699.87=32.84$.
   This gives the value of the term $(Sx^2-(Sx)^2/n)$ and is known as the total sum of squares.
6. Calculate the sums of squares between means
   $=[(54.6)^2/10+(49.2)^2/10+(41.1)^2/10]-C=(298.12+242.06+168.92)$
   $-699.87=9.23$.
7. Calculate the difference between the total sum of squares and the sum of squares between means
   $=32.84-9.23=23.61$.
   This is known as the residual sum of squares.
8. At this stage, it is useful to begin to draw an ANOVA table (Table 2.7). The degrees of freedom for the whole experiment are $(3\times10)-1=29$. There are three groups of tablets and hence three means. There are hence $(3-1)$ 2 degrees of freedom here. Thus, the residual sum of squares has $(29-2)$ 27 degrees of freedom.

TABLE 2.7
Analysis of Variance Table Derived from Tablet Crushing Strength Data in Table 2.6

Source of Error	Sum of Squares	Degrees of Freedom	Mean Square	F
Between means	9.23	2	—	—
Within each group	23.61	27	—	—
Total	32.84	29	—	—

9. The mean squares are obtained by dividing the sum of squares by the relevant number of degrees of freedom. The two mean squares are thus 4.62 and 0.87. These are inserted into Table 2.7.
10. The F ratio (named after Fisher) is the ratio between the mean squares. This equals 5.31 and is inserted into Table 2.7.
11. The ANOVA table is now complete (Table 2.8).
12. The ratio is compared with the appropriate tabulated value of F.

Separate F tables are given in Table A1.3 and Table A1.4 in Appendix 1 for probability levels of 0.05 and 0.01, respectively. Use of either of these tables requires two values for degrees of freedom. That for the mean square between means forms the top row of the table, and that for the mean square of the residuals forms the first column of the table.

For the data under consideration and with a significance level of 0.05, the tabulated value of F is 3.35. Thus, there is a significant difference between the means at $P=0.05$. The corresponding value for F at $P=0.01$ is 5.49. This is greater than the calculated value, and hence, the difference is not significant at that probability level.

The value of ANOVA as a tool should now be apparent. There is no limit to the number of groups of data, and all groups need not necessarily be of the same size.

ANOVA shows that a significant difference occurs between the means of many groups of data. However, it gives no information on which group is significantly different from the others. Therefore, having established that there

TABLE 2.8
Complete Analysis of Variance Table Derived from Tablet Crushing Strength Data in Table 2.6

Source of Error	Sum of Squares	Degrees of Freedom	Mean Square	F
Between means	9.23	2	4.62	5.31
Within each group	23.61	27	0.87	—
Total	32.84	29	—	—

are differences, it is necessary to establish whether all groups differ from each other or some groups are effectively the same. There are many tests available which help establish this point. The simplest of these is to calculate the least significant difference.

2.4.2 THE LEAST SIGNIFICANT DIFFERENCE

This test uses the Student's t value. This is an inappropriate test to use when there are more than two groups of data to *establish* whether significant differences exist. However, it is now used *after* a significant difference has been shown to exist by ANOVA

$$t = \frac{x_{mA} - x_{mB}}{\sqrt{s_p^2 (1/n_A + 1/n_B)}} \tag{2.5}$$

Then, the least significant difference between the means of Batch A and Batch B (i.e., $x_{mA} - x_{mB}$) is

$$t\sqrt{s_p^2 (1/n_A + 1/n_B)}$$

where
   t = the tabulated value of t with the appropriate number of degrees of freedom
      (18) and required significance level (0.05).
In this case, the critical value of t is 2.101.
   The variance s^2 is equal to the mean square within each group (in this case, 0.87). Therefore, the least significant difference

$$= 2.101 \times \sqrt{0.87 \times 2/10} = 0.88$$

The differences between the means are:

Batch A and Batch B: 0.54
Batch A and Batch C: 1.35
Batch B and Batch C: 0.81

Thus, any difference above 0.88 is significant, and in this case the difference between A and C proves significant. Also, though not significant, the difference between B and C approaches 0.88. Hence, this is a reasonable indication that, of the three treatments, Batch C is the one that is most likely to be different.
   There are several other methods of determining which, if any, treatment gives significantly different results after ANOVA. These include the Duncan multiple range test, the Dunnett test, the Tukey multiple range test, and the Scheffé test. All give a parameter equivalent to the least significant difference, and each has its own claimed advantages. Interested readers should refer to a textbook on statistics for further details.

2.4.3 TWO-WAY ANALYSIS OF VARIANCE

The ANOVA test described above is more properly called one-way ANOVA. One factor is deliberately changed (e.g., Batch A, B, or C). However, a situation may arise when two factors are changed. For example, results may be obtained on different equipments or in different geographical areas. The aim is therefore to determine whether the treatments have a significantly different effect when taking the known underlying variation into account. Two-way ANOVA is employed in this case. The situation is best illustrated by a worked example.

In three different countries, a multinational pharmaceutical company produces tablets containing a certain active ingredient. Each country uses its own formulation for the tablets. It is decided to produce the tablets using the same formulation in all the three countries. *In vitro* dissolution data appear to indicate differences among the three formulations, but the differences might occur due to formulations being produced at different sites.

Let the formulations be designated A, B, and C and the three sites of manufacture I, II, and III. Batches are produced at all three sites using all three formulations. Three batches of each formulation are thus obtained, and an ANOVA shows whether significant differences between the batches are present. However, there might be factors connected with the site that affect the results, such as equipment, personnel, or the familiarity a particular site will have with the production of its local formulation. In fact, apparent differences between formulations might be almost entirely due to such factors.

The following experiments are therefore carried out. Tablets of each formulation are prepared at all three sites, and the dissolution of six tablets from each batch is determined. The results are expressed at $t_{50\%}$, the time in minutes for half of the active ingredient contained in each tablet to dissolve. The data are given in Table 2.9. The italicized numbers in the tables are the totals and means for each particular group of six measurements.

The total variance is made up of four components, namely, the variance among formulations, the variance among sites of manufacture, the residual variance, and the variance among determinations within the same group of measurements. The last is termed the within-cell variance. The stages in the calculation of two-way ANOVA are very similar to those in the calculation of a one-way analysis.

1. Calculate the grand total, that is, the sum of all the data, and the totals for each site and for each formulation. These are shown in Table 2.9.
2. Calculate the correction term
   $= (1620)^2/54 = 48,600.$
3. Calculate the total sum of squares and subtract the correction term
   $= [(33)^2 + (37)^2 + \cdots + (16^2)] - 48,600 = 3258.$
4. Calculate the between formulations sum of squares
   $= [(708)^2/18 + (504)^2/18 + (408)^2/18] - 48,600 = 2608.$
5. Calculate the between sites sum of squares
   $= [(504)^2/18 + (576)^2/18 + (540)^2/18] - 48,600 = 144.$
6. Because each cell contains replicated results, there is an additional stage in the calculation, that of the within-cell sum of squares. If measurements

TABLE 2.9
Dissolution Data ($t_{50\%}$ Minutes) from Tablets Made from Three Formulations (A, B, and C) at Three Sites (I, II, and III)

Site	Formulation									Site total
	A				B			C		
I	33	37	35	22	24	30	23	23	25	
	36	33	36	28	29	23	21	24	22	
	210				*156*			*138*		504
	35				*26*			*23*		
II	41	38	39	27	27	29	23	24	26	
	42	44	42	33	31	27	28	27	28	
	246				*174*			*156*		576
	41				*29*			*26*		
III	42	38	39	28	27	32	19	19	19	
	42	42	49	33	29	25	20	21	16	
	252				*174*			*114*		540
	42				*29*			*19*		
Formulation total	708				504			408		1620

The italicized figures are the totals and means for each group of six determinations.

had not been replicated, this step would not be carried out. The mean of each cell is subtracted from every result in that cell and the difference squared. Thus, the within-cell sum of squares is
$$=(33-35)^2+(37-35)^2+\cdots+(16-19)^2=294.$$
7. The residual sum of squares
$$=3258-(144+2608+294)=212.$$
8. The ANOVA table can now be constructed (Table 2.10). The degrees of freedom are calculated as follows. The total number of degrees of freedom for n observations is $n-1$, in this case 53. If there are R rows in the table and C columns, then the numbers of degrees of freedom associated with rows and columns are $R-1$ and $C-1$, respectively. In this case, there are 2 degrees of freedom associated with both. The degrees of freedom

TABLE 2.10
Analysis of Variance Table for Dissolution Data from Table 2.9

Sources of Error	Sum of Squares	Degrees of Freedom	Mean Square	F
Between formulations	2608	2	1304.0	200.6
Between sites	144	2	72.0	11.1
Residuals (interaction)	212	4	53.0	8.2
Within cells	294	45	6.5	—
Total	3258	53	—	—

associated with the residuals are $(R-1)\times(C-1)$, in this case 4. Degrees of freedom associated with the error within the cells thus total 45.
9. F is calculated by dividing the mean squares for formulation, site, and residuals by the mean square for within cells.

Tabulated values of F from Table A1.3 in Appendix 1 at $P=0.05$ are $F_{2.45}=3.21$ and $F_{4.45}=2.59$. All effects are thus significant at this level of probability, though the effect of the formulation is much greater than that of the others. The residual term is called the interaction term. In the absence of interaction, the interaction mean square would, on average, equal the within-cell mean square.

FURTHER READING

Bolton, S. and Bon, C., *Pharmaceutical Statistics: Practical and Clinical Applications*, 4th ed., Marcel Dekker, New York, 2004.
Clarke, G.M. and Cooke, D., *A Basic Course in Statistics*, 4th ed., Arnold, London, 1998.
Jones, D. S., *Pharmaceutical Statistics*, Pharmaceutical Press, London, 2002.

REFERENCE

1. *United States Pharmacopoeia*, 28th ed., Chapter 1010, United States Pharmacopeial Convention, Rockville, MD, 2005.

3 Nonparametric Methods

3.1 INTRODUCTION

The tests so far employed for comparing the means of groups of data (*t*-test and analysis of variance [ANOVA]) depend on the assumption that the populations involved are normally distributed. In many cases, this cannot be known with certainty, though it can often be assumed. Moreover, the distribution of a sample mean approaches that of a normal distribution as the sample size is increased. However, increase in the sample size may not be practicable. A further consideration are that the data to be manipulated by parametric methods must have numerical values. Ordinal data based on rank order, for example, social class or severity of reaction, are not amenable to parametric treatment. However, there is a series of nonparametric tests available that are designed to handle such information. These have the advantage that they make no prior assumptions about the underlying distribution and parameters of the population involved.

As in parametric tests of comparison, the distinction must be made whether the two samples come from independent populations or whether the variates are paired in some way, perhaps by each subject acting as its own control. This obviously depends on the design of the experiment. Hence, here is another example of the design of the experiment and the method of evaluating the results being inextricably linked. Some of the tests which can be used are:

1. Sign test for paired data.
2. Wilcoxon signed rank test for paired data (the Mann–Whitney U-test is very similar).
3. The Wilcoxon two-sample test for unpaired data.

3.2 NONPARAMETRIC TESTS FOR PAIRED DATA

3.2.1 THE SIGN TEST

This is used to test the significance of the difference between the means of two sets of data in a paired experiment. Each subject thus acts as its own control. Only the sign of the differences between each pair of data points is used, and because of its simplicity, this test may be used for a rapid examination of data before a more sensitive test is applied.

Consider the following example. The dissolution rate of tablets is measured on a long-established piece of apparatus. Some modifications to the apparatus have been proposed, but it has been suggested that these will alter the dissolution rate of

TABLE 3.1
Percentage of the Active Ingredient of Ten Tablet Formulations (A to L)
Dissolved for 30 min, Using Two Different Pieces of Dissolution Apparatus
(I and II)

Tablet Formulation	Apparatus I	Apparatus II	Difference (II − I)
A	83	88	+5
B	59	66	+7
C	78	83	+5
D	79	79	0
E	88	92	+4
F	82	90	+8
G	90	92	+2
H	81	83	+2
I	87	77	−10
J	65	68	+3
K	68	72	+4
L	83	89	+6

tablets tested in it. The dissolution rates of ten different tablet formulations (A to L) are measured both on the old apparatus (I) and on the modified apparatus (II), giving the results in Table 3.1.

There is no suggestion that the new apparatus will cause tablets to dissolve either more quickly or more slowly; therefore, this is a two-tail test. The ten formulations are all different, so there is no reason why the data should be normally distributed. Hence, a nonparametric test should be used.

Tabulation of the differences between Apparatus I and Apparatus II shows ten positive signs and one negative sign, and in one case, both pieces of apparatus give the same result. For the purposes of these calculations, results that are tied are ignored. Hence, data from Formulation D are omitted, leaving the data from 11 formulations to be considered.

If the two pieces of apparatus were truly equivalent, then the probability of either a positive or a negative result for any given formulation would be 0.5. When the number of observations is small, the probabilities of various experimental outcomes, that is, 11 positives, 10 positives, and 1 negative etc, can be calculated from the binomial distribution. The number of positive or negative signs needed for significance for the sign test is given in Table 3.2.

For 11 pairs of observations, there should be at least 10 with the same sign for a significant difference at the 5% level, and so the two pieces of dissolution apparatus appear to differ significantly at this level. For a significant difference at the 1% level, all 11 pairs should have the same sign.

For a larger number of observations, (3.1) can be used.

$$Z = \frac{\left|\text{number of positive signs} - \text{number of negative signs}\right| - 0.5}{\sqrt{(\text{number of positive signs} + \text{number of negative signs})}} \quad (3.1)$$

TABLE 3.2
Number of Positive or Negative Signs Needed for
Significance for the Sign Test

Sample Size	Number of Positive or Negative Signs for Significance At	
	5% Level	1% Level
6	6	—
7	7	—
8	8	8
9	8	9
10	9	10
11	10	11
12	10	11
13	11	12
14	12	13
15	12	13
16	13	14
17	13	15
18	14	15
19	15	16
20	15	17

If Z is greater than 1.96, there is a significant difference at the 5% level of probability, and if Z is greater than 2.60, there is a significant difference at the 1% level. Substitution of the numbers in the last row of Table 3.2 into (3.1) gives values for Z of 2.12 and 3.02, respectively, for the two levels of significance.

An important aspect of experimental design arises from the information in Table 3.2. For a 5% level of significance, the smallest number of paired observations that can be expected to yield a significant result is six, and for a 1% level, the corresponding number is eight. If this test is to be used, planning of the experiment must take these minimum requirements into account, and sufficient replicate determinations must be made.

3.2.2 THE WILCOXON SIGNED RANK TEST

This is a more sensitive nonparametric test. In this, the magnitude of the difference between the paired variates as well as its sign is taken into account.

From the data given in Table 3.1, the differences are ranked in order of increasing magnitude, disregarding the sign. Ties such as D are discounted, and identical differences are given a mean rank. Thus, G and H have a rank of $(1+2)/2 = +1.5$ (Table 3.3a). The results are then rearranged taking into account the signs and their magnitude. Ranks with negative signs and ranks with positive signs are summed separately (Table 3.3b).

TABLE 3.3A
Assigned Ranks with and without Signs for
Data from Table 3.1, Arranged in Increasing
Order of Magnitude

Formulation	Assigned Rank	Rank with Sign
G	1.5	+1.5
H	1.5	+1.5
J	3	+3
E	4.5	+4.5
K	4.5	+4.5
A	6.5	+6.5
C	6.5	+6.5
L	8	+8
B	9	+9
F	10	+10
I	11	−11

TABLE 3.3B
Ranks with Positive and Negative Signs
Derived from Table 3.3a

	Ranks	
	Positive Signs	Negative Signs
	+1.5	−11
	+1.5	
	+3	
	+4.5	
	+4.5	
	+6.5	
	+6.5	
	+8	
	+9	
	+10	
Sum	+55	−11

Table 3.4 gives the values of the smaller of the two rank sums at a 5% significance level for a range of sample sizes. The smaller rank sum must be equal to or less than the number given in the table. Significance is established neither at the 5% level nor at the 1% level.

TABLE 3.4
Values Giving Significance for the
Wilcoxon Signed Rank Test

Sample Size	Significance Level	
	5%	1%
6	0	—
7	2	—
8	3	0
9	5	1
10	8	3
11	10	5
12	13	7
13	17	10
14	21	13
15	25	16
16	30	19
17	35	23
18	40	28
19	46	32
20	52	37

3.3 NONPARAMETRIC TESTS FOR UNPAIRED DATA

3.3.1 THE WILCOXON TWO-SAMPLE TEST

This test deals with two groups of data that have been obtained independently. An item in one group does not act as the control for an item in the second group. The data need not be normally distributed, and the groups need not even be of the same size. The test is best illustrated by an example.

Armstrong et al.[1] described a method for measuring the release of drugs from oily bases. The drug was released into an aqueous medium, which was then analyzed. The method was used to compare drug release from bases of differing composition, and some of the data obtained are given in Table 3.5.

The objective is to answer the question: Does a change of base have a significant effect on drug release? Cursory examination of the data indicates that Base B gives a slower release, in that most of the values for Base B are less than those of Base A. However, no suggestion is made that release is either specifically accelerated or specifically retarded; hence, a two-tail test is indicated. There is no evidence that the data are normally distributed; otherwise, a t-test could have been used.

The Wilcoxon two-sample test is carried out by arranging the data in ascending order of magnitude (Table 3.6).

TABLE 3.5
Drug Release from Two Topical Bases after
120 min (Data are mg% in the Aqueous Phase)

	Sample Size	
	Base A	Base B
	0.782	0.742
	0.790	0.779
	0.798	0.748
	0.772	0.764
	0.790	0.757
Mean	0.786	0.758

TABLE 3.6
Drug Release from Two Topical Bases after 120 min

	Base A	Rank	Base B	Rank
	—	—	0.742	1
	—	—	0.748	2
	—	—	0.757	3
	—	—	0.764	4
	0.772	5	—	—
	—	—	0.779	6
	0.782	7	—	—
	0.790	8.5	—	—
	0.790	8.5	—	—
	0.798	10	—	—
Sum of ranks		39		16

Note: The results shown in this table are arranged in rank order (data are mg% in the aqueous phase).

The sum of the ranks of data from Base B is

$$1+2+3+4+6=16$$

Similarly, for Base A, the sum is

$$5+7+8.5+8.5+10=39$$

Adding 16 to 39 gives 55, which is the sum of the integers 1 to 10. This has no bearing on the outcome of the experiment, but it serves as a useful check whether the ordering has been carried out correctly. Note that in Group A there are two identical results (0.790). If these were slightly different, they would be ranked 8 and 9 in the ascending order. These positions are therefore averaged (8.5), and this rank is given to each. The total remains the same.

If there were no difference in drug release between the two bases, then the totals for each group would be about the same. The difference (16 to 39) looks large and is an indication of a difference in drug release, but nevertheless could have occurred by chance.

The next step is to determine how many of all the possible arrangements of five of the numbers 1 to 10 will give a total of 16 or less. There are two, namely,

$$1+2+3+4+5=15$$

$$1+2+3+4+6=16$$

If r objects are taken from a total of n objects, then the number of different combinations is given by $n!/[r!(n-r)!]$. In this example, $n=10$ and $r=5$, so 252 different combinations are possible. Thus, the probability (P) of obtaining a sum of ranks less than or equal to 16 is $2/252=0.00794$.

P, as calculated above, is for a one-tail test, in that it is the probability that release from Base B is less than that from Base A. If a significant difference between Base A and Base B is to be established, then P should be doubled and becomes 0.0159. From the above, it follows that a significant difference occurs between the two bases at a probability level of 1.59%. It is of interest to apply the t-test to these data, assuming for the moment that both populations follow a normal distribution. t is found to be 3.63, which is equivalent to a probability of 1.66% for a two-tail test, close to that calculated by the Wilcoxon test.

This test, being nonparametric, can also be applied to nonnumerical data. Consider the following example. A new treatment has been devised for patients suffering from a particular disease. In a group of eight patients, four patients receive the new treatment (designated N) and four the old treatment (designated O). A double-blind trial is carried out to protect from bias, and improvements in the condition of the patients are assessed by an independent observer. The observer puts the degree of improvement of the patients into ascending rank order, giving the information in Table 3.7. Thus, the two patients who showed the greatest improvement had received the new treatment, and the old treatment had been given to the two who showed the least improvement.

The sum of ranks of patients receiving the new treatment is

$$3+5+7+8=23$$

If there were no difference between the two treatments, then the ranks 1 to 8 would be assigned at random to the patients. Therefore, the ranks scored by the four patients who were on the new treatment would be any combination of four from the

TABLE 3.7
Rank Order of Improvement of Patients Receiving Either Old (O) or New (N) Treatments

Improvement (Rank Order)	1	2	3	4	5	6	7	8
Treatment received	O	O	N	O	N	O	N	N

TABLE 3.8
**Possible Combinations of Four Items Chosen from a Group
of Eight, Giving Their Rank Sums**

				Ranks				
1	2	3	4	5	6	7	8	Rank Sum
N	N	N	N					10
N	N	N		N				11
N	N		N	N	N			12
N	N	N						12
N		N	N	N				13
—	—	—	—	—	—	—	—	—
			N		N		N	N	23
	N					N	N	N	23
			N	N	N			N	23
	N				N	N	N	24
			N	N			N	N	24
		N			N	N	N	25
			N	N	N	N	26

numbers 1 to 8. There are 70 ways in which four items can be selected from a group of eight, and all are equally likely. Some of these combinations are shown in Table 3.8.

There is 1 chance in 70 of getting a rank sum of 26, 1 chance in 70 of a rank sum of 25, 2 in 70 of a sum of 24, and 3 in 70 of a sum of 23. Therefore, the total probability of achieving a rank sum greater than or equal to 23 is $(3+2+1+1)/70=0.1$. Because a significant improvement rather than a significant difference is being sought, this is a one-tail test. Thus, if a probability of 0.05 is required as an indication of success, this experiment must be considered a failure.

A key point for all nonparametric tests is that an adequate number of replicate determinations are carried out. In the last example, if only four patients had been used, two on the new treatment and two on the old, then it would have been impossible to achieve a significance of 0.05, even if the two patients receiving the new treatment had shown the most improvement. If six patients had been treated, then the new treatment would have had to achieve ranks 4, 5, and 6 for the result to be significant at this level of probability. Any other combination of results would not have established a significant difference.

FURTHER READING

Clarke, G. M. and Cooke, D., *A Basic Course in Statistics*, 4th ed., Arnold, London, 1998.

REFERENCE

1. Armstrong, N. A., Griffiths, H.-A., and James, K. C., An *in-vitro* model to simulate drug release from oily media, *Int. J. Pharm.*, 41, 115, 1988.

4 Regression and Correlation

4.1 INTRODUCTION

Many experiments consist of changing the value of a factor (the independent variable, the predictor) and measuring the response (the dependent variable, the outcome). This produces many pairs of data points. It is often convenient to present these in graphical form, and it is conventional to plot the factor on the X axis (the abscissa) and the response on the Y axis (the ordinate). The independent variable, for example, time and concentration, is chosen by the experimenter and should be subject to negligible error. The response is subject to random error associated with measurement.

Before the advent of computers, a rectilinear relationship was detected by plotting a graph of the response against the factor and observing whether the points could reasonably be considered to follow a straight line. The best straight line was judged subjectively, and if required, the slope and intercept of that line were obtained by measuring distances on the graph. A method for calculating the best equation relating the points, called regression or least squares analysis, was known, but it was a highly protracted procedure, particularly if there were many data points. Though such calculations are now carried out easily by a computer, the results they yield can, without careful consideration, lead to inappropriate conclusions.

Regression is the process of deriving a relationship between one or more factors and a response. When a factor and a response are directly related, a plot of one against the other will be a straight line. Hence, linear regression applies to such relationships.

The closely related topic of correlation seeks to determine how well a linear or other equation describes the relationship between the variables. Both regression and correlation are important techniques in experimental design and are particularly relevant to model-dependent designs and to constructing and interpreting contour plots, topics that are dealt with in later chapters. It is thus worthwhile to consider both topics in some detail. For deriving many of the expressions used in this chapter, the reader is referred to texts on statistics such as *A Basic Course in Statistics* by Clarke and Cooke[1] and *An Introduction to Linear Regression and Correlation* by Edwards.[2]

4.2 LINEAR REGRESSION

The process of linear regression can be illustrated using data gathered by Gebre-Mariam et al.[3] These workers were interested in the diffusion of solutes through glycerogelatin gels such as the shells of soft capsules. In such systems, the gelatin forms a porous matrix, and the pores are filled with a mixture of glycerol and water. As with all diffusion processes, the viscosity of the liquid medium was thought to play an important role. With this in mind, the viscosities of a series of mixtures of glycerol and water were determined. The data are given in Table 4.1 and shown graphically in Figure 4.1.

The equation of the best-fitting straight line through the data points is called the regression line and can be calculated by regression or least squares analysis. It takes the form shown in (4.1):

$$y = b_0 + b_1 x \tag{4.1}$$

where

b_1 = the slope of the line
b_0 = the intercept on the ordinate, that is, the value of y when $x = 0$.

Information used for calculating b_0 and b_1, derived from data given in Table 4.1, is given in Table 4.2.

TABLE 4.1
Viscosities of Mixtures of Glycerol and Water at 23 °C

Glycerol (%w/w)	12.3	18.5	24.6	30.8	36.9
Viscosity ($N \cdot s \cdot m^{-2} \times 10^3$)	4.83	6.32	7.50	9.66	11.90

Source: Gebre-Mariam et al.[3]

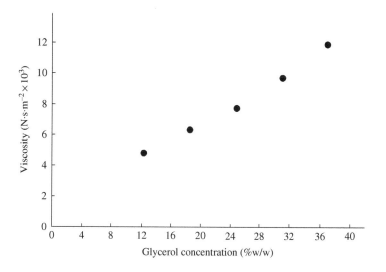

FIGURE 4.1 Viscosities of mixtures of glycerol and water at 23 °C (Gebre-Mariam et al.[3]) (data taken from Table 4.1).

TABLE 4.2
Viscosities of Mixtures of Glycerol and Water at 23 °C and Derived Information Used in the Calculations

Glycerol (%w/w) (x)	Viscosity (N·s·m^{-2} × 10^3) (y)	x^2	y^2	xy
12.3	4.83	151.3	23.32	59.41
18.5	6.32	342.3	39.94	116.92
24.6	7.50	605.2	56.25	184.50
30.8	9.66	948.6	93.32	297.53
36.9	11.90	1361.6	141.61	439.11

n	5	5			
Sum	123.1	40.21	3409.0	354.44	1097.47
Sum^2	15,153.6	1616.84	11,621,281	125,627.71	
Mean	24.6	8.04			

Source: Gebre-Mariam et al.[3]

The regression line is the line for which the sum of the vertical distances between it and the experimental points is less than the sum obtained with any other straight line. The vertical distances are known as residuals. If the observed result is Y_{obs} and the value predicted by the regression line is Y_{pred}, then the residual is $Y_{obs} - Y_{pred}$. Some residuals will be positive and some negative and so will tend to cancel each other out. This is avoided by squaring the residuals so they all become positive before summation. This is the origin of the phrase "least squares analysis."

The slope (b_1) of the regression line, known as the regression coefficient, is calculated from (4.2)

$$b_1 = \frac{\sum(xy) - \sum(x)\sum(y)/n}{\sum(x^2) - \sum{}^2(x)/n} \qquad (4.2)$$

$\sum(xy)$ is the sum of the products of each value of x and the corresponding value of y, so that for the information given in Table 4.1

$$\sum(xy) = (12.3 \times 4.83) + (18.5 \times 6.32) + \cdots + (36.9 \times 11.90) = 1097.47$$

$\sum(x)\sum(y)/n$ is the sum of all the x multiplied by the sum of all y, divided by the number of pairs of data points. Because $\sum(x) = 123.1$ and $\sum(y) = 40.21$, then

$$\frac{\sum(x)\sum(y)}{n} = \frac{123.1 \times 40.21}{5} = 989.97$$

$\sum(x^2)$ is the sum of the square of each value of x; therefore,

$$\sum(x^2) = 12.3^2 + 18.5^2 + \cdots + 36.9^2 = 3408.95$$

$\Sigma^2(x)/n$ is the squared sum of all the x's, divided by the number of pairs of x and y. Therefore,

$$\frac{\Sigma^2(x)}{n} = \frac{123.1^2}{5} = 3030.72$$

Substitution into (4.2) gives b_1

$$b_1 = \frac{1097.47 - 989.97}{3408.95 - 3030.72} = 0.284$$

Substituting 0.284 for b_1, together with the mean value for x (24.6) and the mean for y (8.04) into (4.1) yields the regression equation (4.3)

$$y = 1.045 + 0.284x \qquad (4.3)$$

This is the regression equation of Y on X, because Y, the response, is estimated from X, the factor. It is represented by the continuous line in Figure 4.2.

By substituting any value of x between 12.3% glycerol and 36.9% glycerol into (4.3), the viscosity of a mixture of that strength can be confidently predicted. However, for mixtures containing less than 12.3% glycerol or more than 36.9% glycerol, prediction of viscosity involves extrapolation, that is, using concentrations outside the range that has been explored experimentally. Such predictions must always be made with caution, because no evidence has been obtained to suggest that (4.3) applies outside the concentration range that has been studied. Extrapolation is particularly dangerous for relationships that are not rectilinear.

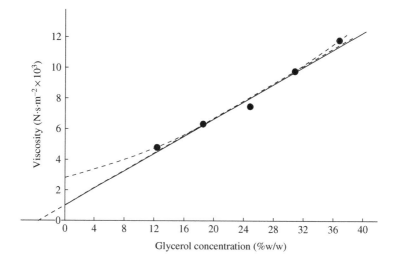

FIGURE 4.2 The relationship between viscosity (Y) and glycerol concentration (X). Continuous line: the linear regression line of Y on X; dashed line: the linear regression line of X on Y; dotted line: the quadratic regression line of Y on X.

The general equation for the regression line is given by (4.4).

$$Y = b_0 + b_1 X_1 + b_2 X_2 + \cdots + b_{n-1} X_{n-1} + b_n X_n \tag{4.4}$$

where
   Y = the response
   X_1 to X_n = n factors
   b_0 to b_n = the coefficients.

If there is only one factor and one response, $n = 1$, and so (4.4) reduces to (4.1). Calculation of the coefficients of a regression equation is now conveniently carried out by a computer. Most suites of statistical programs will give not only these coefficients but also considerable information on how well the regression equation describes the relationship between the independent variables and the response, that is, the goodness of fit. The output of statistical analysis by Microsoft Excel® will be used as an example. In Microsoft Excel, use of the LINEST command calculates the statistics of a line using the least squares method and returns an array of numbers which describes that line. These are shown in Table 4.3.

Regression of the data in Table 4.1 gives the output shown in Table 4.4. The significance of this information is now discussed.

4.2.1 THE NUMBER OF DEGREES OF FREEDOM (CELL B11 IN TABLE 4.4)

This is governed by the number of pairs of data points used to calculate the regression equation. Obviously, the greater the number of pairs of data points, the more reliable will be the equation as a means of predicting new information. The number of degrees of freedom is given by (4.5).

$$\text{degrees of freedom} = n - (k + 1) \tag{4.5}$$

where
   n = the number of pairs of data points (in this case 5)
   k = the number of factors in the regression equation (in this case 1).

Therefore, the number of degrees of freedom in this case is 3.

TABLE 4.3
Microsoft Excel Output after Linear Regression Using the LINEST Command

Coefficient b_n	Coefficient b_{n-1}	...	Coefficient b_1	Coefficient b_0
Standard error of b_n	Standard error of b_{n-1}		Standard error of b_1	Standard error of b_0
Coefficient of determination (r^2)	Standard error of Y		#N/A	#N/A
F	Degrees of freedom		#N/A	#N/A
Regression sum of squares	Residual sum of squares		#N/A	#N/A

TABLE 4.4
Microsoft Excel Output after Linear Regression of the Data Given in Table 4.1

	A	B
1	Glycerol concentration	Viscosity
2	12.3	4.83
3	18.5	6.32
4	24.6	7.50
5	30.8	9.66
6	36.9	11.90
7		
8	0.284 (coefficient b_1)	1.045 (coefficient b_0)
9	0.022 (standard error of b_1)	0.562 (standard error of b_0)
10	0.983 (coefficient of determination)	0.419 (standard error of Y)
11	174.17 (F)	3 (degrees of freedom)
12	30.55 (regression sum of squares)	0.526 (residual sum of squares)

4.2.2 THE COEFFICIENT OF DETERMINATION (r^2) (CELL A10 IN TABLE 4.4)

The coefficient of determination is the square of the correlation coefficient (r).

The basis of the correlation coefficient can be seen if two straight lines are drawn at right angles and parallel to the axes of Figure 4.1, intersecting at the mean values of x and y ($x_m = 24.6$ and $y_m = 8.04$, respectively). If x is positively related to y, then most points will be located in areas B and C, and if they are negatively related, most points will be in areas A and D. If they are unrelated, the points will be scattered in all four areas (Figure 4.3).

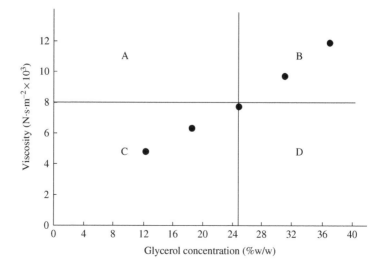

FIGURE 4.3 The graphical basis of the linear correlation coefficient.

For all points in areas B and C, the terms $x-x_m$ and $y-y_m$ will both be either positive or negative, whereas in areas A and D they will have different signs. Therefore, for positive relationships between x and y, $\Sigma(x-x_m)(y-y_m)$ will have a positive value, and for negative relationships it will have a negative value. If the points were to be distributed over all four areas, those in A and D will tend to cancel out those in B and C, so that the correlation coefficient will have a low value, which might be positive or negative.

The value of the correlation coefficient ranges from -1 through 0 to $+1$. However, it is conventional for the sign not to be quoted, and so it is reported in the range 0 to 1.

The correlation coefficient is given by (4.6)

$$r = \frac{\sum(xy) - \sum(x)\sum(y)/n}{\sqrt{\left[\sum(x^2) - \sum{}^2(x)/n\right]\left[\sum(y^2) - \sum{}^2(y)/n\right]}} \qquad (4.6)$$

Substituting the appropriate values from Table 4.2 gives

$$r = \frac{1097.47 - 989.97}{\sqrt{[(3408.95 - 3030.72)(354.45 - 323.37)]}} = 0.99$$

The higher the value of r, the greater the likelihood that x and y are correlated. What constitutes a satisfactory value of r depends on the value of n, the number of pairs of data points. If $n=2$, r must have the maximum possible value of 1.000, no matter what the data are. If n is a larger number, a lower value of r would be acceptable as evidence of correlation. Table 4.5 shows values of the correlation coefficient for a range of sample sizes that are significant at a 5% probability level. There are three

TABLE 4.5
Theoretical Values of the Correlation Coefficient
$(P=0.05)$

Number of Pairs of Data Points (n)	Degrees of Freedom (ϕ)	Correlation Coefficient (r)
3	1	0.997
4	2	0.950
5	3	0.878
6	4	0.811
7	5	0.754
8	6	0.707
9	7	0.666
10	8	0.632
15	13	0.514
20	18	0.444
30	28	0.361
50	48	0.279
100	98	0.197

degrees of freedom ($\Phi=3$), and so the theoretical value of r is 0.878. As this is lower than the calculated value of r, it is likely that x and y are correlated.

The coefficient of determination is the proportion of the variation of the dependent variable that is explained by the regression equation. In regression analysis, the squared difference between the observed value of Y and that predicted by the regression equation is calculated. The total of these squared differences is called the residual sum of squares (cell B12 in Table 4.4). The difference between each observed value of Y and the mean value of Y is also calculated. After squaring and summation, the total sum of squares is obtained. This in turn is equal to the regression sum of squares (cell A12 in Table 4.4) plus the residual sum of squares. The smaller the residual sum of squares is when compared with the total sum of squares, the better the equation explains the relationship between the variables.

Thus, for the data shown in Table 4.4, the regression sum of squares (cell A12) is 30.55 and the residual sum of squares (cell B12) 0.526. Therefore, the total sum of squares is 31.08, and the ratio between the regression sum of squares and the total sum of squares

$$= 30.55/31.08$$
$$= 0.983 = r^2 \text{ (cell A10)}$$

Thus, 98.3% of the variation in the viscosities of the glycerol solutions with concentration is explained by (4.3).

4.2.3 THE STANDARD ERRORS OF THE COEFFICIENTS (CELLS A9 AND B9 IN TABLE 4.4)

The value of the standard error indicates that if the experiment were to be repeated, the value of the coefficient b_1 should lie between 0.284 ± 0.022. Also, the value of the coefficient b_0 should be between 1.045 ± 0.562. The greater the standard error of the coefficient, the less reliable is the coefficient and the less likely that the regression equation represents the raw data.

The confidence in a coefficient can be assessed by dividing the coefficient by its standard error and comparing the result with the appropriate tabulated value of Student's t. Taking coefficient b_1 as an example, the t value of the observed results is $0.284/0.022 = 12.91$. There are three degrees of freedom. The critical t value at a probability level of 0.05 is 2.35, so the chance that the coefficient does not represent a true relationship is less than 1 in 20. Thus, the glycerol concentration is an important factor in predicting viscosity.

4.2.4 THE F VALUE OR VARIANCE RATIO (CELL A11 IN TABLE 4.4)

The F value indicates whether the equation is a true relationship between the results rather than coincidence, using the analysis of variance technique described in Chapter 2. The calculated value of F (in this case, 174.17) is compared with the tabulated critical values of F shown in Table A1.3 and Table A1.4. The numbers running along the top of Table A1.3 and Table A1.4 represent the number of variables on the right-hand side of the regression equation (in this case, 1 — the glycerol concentration), and the numbers running down the left of the table represent the number of degrees of freedom. The critical value of $F_{1.3}$ is 34.1 at a probability level of 0.01; therefore, the probability of the relationship being due to chance is less than 1 in 100.

4.2.5 THE TWO REGRESSION LINES

The reason why Gebre-Mariam et al.[3] collected the data shown in Table 4.1 was to predict the viscosity of the contents of the pores in a glycerogelatin matrix. Thus, glycerol concentration was the independent variable (X) and viscosity the dependent variable (Y). The regression line was that of Y on X, and the residuals were the vertical distances of the regression line from the data points, that is, $y_{obs} - y_{pred}$.

As this work progressed, it was found that the viscosity of the fluid within the pores could be measured by electron spin resonance. This in turn could be used to predict the composition of the fluid, which might not have the same composition as the mixture of glycerol and water originally used to make the gel. Now the viscosity is the independent variable, glycerol concentration is the dependent variable, the regression line is of X on Y, and the residuals are the horizontal distances from the regression line to the data points ($x_{obs} - x_{pred}$).

If the regression line passes perfectly through all the points, the correlation coefficient will equal 1 and the two regression lines will be the same. This cannot be the case with data in Table 4.1, as it has been already shown that the correlation coefficient is less than 1.

The slope of the regression line of X on Y is given by (4.7) and equals 3.46.

$$b_1 = \frac{\sum(xy) - \sum(x)\sum(y)/n}{\sum(y^2) - \sum{}^2(y)/n} \qquad (4.7)$$

The intercept on the X axis is −3.20. Therefore, the regression equation of X on Y is

$$x = -3.20 + 3.46y$$

This is shown as a dashed line in Figure 4.2.

4.3 CURVE FITTING OF NONLINEAR RELATIONSHIPS

There is an infinite number of ways in which a pair of variables may be related. The simplest situation occurs when the variables are directly proportional, so that a plot of one variable against the other yields a straight line. Linear regression analysis can then be applied. However, variables may not be directly proportional but are otherwise related, and alternative means must be applied to derive a mathematical formula that fits the results.

4.3.1 THE POWER SERIES

Often, a plot of one variable against another follows a regular profile that is not a straight line. Fitting the results to a power series, which is a series of terms in progressively increasing or decreasing powers of the independent variable, is an empirical method of obtaining an equation that relates such data.

The virial equation for unexpanded gases is probably the best-known example of the power series approach. Expanded gases follow the perfect gas law (4.8)

$$\frac{PV}{RT} = 1 \tag{4.8}$$

where
  P, V, and T = pressure, volume, and temperature, respectively
  R = the gas constant.

As the volume decreases, gas molecules move closer together and intermolecular forces become more important. As a result, the volume of the gas deviates more and more with increasing pressure from the relationship expressed in (4.8). Deviations are accommodated in the virial equation by transforming the relationship to a power series in V, as shown in (4.9)

$$\frac{PV}{RT} = 1 + BV^{-1} + CV^{-2} + DV^{-3} + \cdots \tag{4.9}$$

where
  B, C, and D = virial coefficients and are constant for the system under scrutiny.

The powers of V are negative, so that as new powers of V are added, the correction becomes progressively smaller and usually ceases to be of practical importance after the third term on the right-hand side. Therefore, as V increases, the series progressively reduces to (4.8).

Tests for goodness of fit of polynomial equations present the problem that linearity cannot be established by plotting a function of one variable against a function of the other, because the terms on the right-hand side of the equation cannot be resolved into one function of the independent variable. The easiest way of testing such a relationship is to compare observed results with calculated results in the form of a table or a plot or through regression analysis of observed results against calculated results. For a good fit, a straight line passing through the origin should be observed.

4.3.2 QUADRATIC RELATIONSHIPS

The quadratic equation, of which (4.10) is an example, is the lowest power series and is represented graphically by a parabola.

$$y = b_0 + b_1 x + b_2 x^2 \tag{4.10}$$

The terms making up (4.10) control the shape and position of the resulting parabola. Thus

1. When b_2 is positive, the parabola passes through a minimum, and when b_2 is negative, it passes though a maximum.
2. If $b_0=b_1=0$, the plot will be symmetrical about the Y axis, with the minimum or maximum, whichever applies, passing through the origin.
3. If $b_1=0$ but $b_0 \neq 0$, the plot will still be symmetrical about the Y axis, but the minimum or maximum will pass through $x=0$ and $y=b_0$.
4. If neither b_0 nor $b_1=0$, the plot will not be symmetrical about the Y axis, and the minimum or maximum will pass through $b_1/2b_2$.
5. As b_1 increases with b_2 remaining constant, the arms become steeper and the parabola narrows. Alternatively, as b_2 decreases with b_1 remaining constant, the arms become less steeper and the parabola broadens.

The position of the maximum or minimum of a parabola is easily determined by differentiating the equation and placing the result equal to zero.

Because the regression lines and correlation coefficients are now so easy to calculate, it may be tempting to attempt to fit data to a parabola to ascertain whether an improved correlation coefficient is obtained compared with linear regression. For example, fitting the data given in Table 4.1 to a quadratic equation yields (4.11)

$$\text{viscosity} = 2.35 + 0.139C + 0.00322C^2 \tag{4.11}$$

where
   $C=$ the concentration of glycerol.

This relationship is shown graphically as the dotted line on Figure 4.2. The correlation coefficient is 0.998, which is slightly greater than the linear correlation coefficient calculated from (4.3). However, such an approach needs careful consideration. The curvature of the dotted line in Figure 4.3 indicates that at the extreme values of x, the values of the residuals, that is, $y_{obs} - y_{pred}$, are at their highest. Hence, using this line to predict viscosity at low and high glycerol concentrations will give predictions of ever-increasing error. This emphasizes the danger of extrapolating from nonlinear relationships to which reference has been already made.

4.3.3 CUBIC EQUATIONS

When a plot of y against x that deviates from linearity fails to fit a quadratic equation, the next step up the power series may be considered. This is the cubic equation as shown in (4.12)

$$y = b_0 + b_1 x + b_2 x^2 + b_3 x^3 \tag{4.12}$$

Theoretically, cubic equations should have three solutions and pass through a maximum and a minimum, but in practice, this is not always the case. As with quadratic equations, the shapes of the plots depend on the signs and magnitudes of the coefficients.

4.3.4 Transformations

Rather than fit a polynomial to curved data, it is often preferable to try transformations to determine whether a simpler model can be found. Rectilinear relationships
can sometimes be found by plotting the logarithms of one or both variables.

Thus, an exponential relationship (4.13) can be transformed into a straight-line
relationship (4.14).

$$y = b_0 b_1^x \tag{4.13}$$

$$\log y = \log b_0 + x \log b_1 \tag{4.14}$$

Similarly, a geometric relationship (4.15) can be expressed in logarithmic form (4.16)

$$y = b_0 x^{b_1} \tag{4.15}$$

$$\log y = \log b_0 + b_1 \log x \tag{4.16}$$

If a plot of y against x gives a line that curves upward at high values of x, it is
necessary to shrink the upper end of the y scale. This can be achieved by plotting
$y^{1/2}$ or $\log y$ or $-1/y$ against x. All these have the required effect on the y scale, $-1/y$
having the greatest impact. If the line curves downward, the upper end of the x scale
should be shrunk. Therefore, y can be plotted against either $x^{1/2}$, $\log x$, or $-1/x$. It is
also possible to transform both scales.

4.4 MULTIPLE REGRESSION ANALYSIS

Regression and correlation, as discussed so far, concern the relationship between
one dependent variable (y) and one independent variable (x). Because there are only
two variables, the relationship involves only two dimensions, so that the results can
be plotted as a line that may or may not be straight.

Sometimes, more than two variables are involved. A dependent variable may
be related to two independent variables x_1 and x_2, as in (4.17). b_0, b_1, and b_2 are
constants or coefficients.

$$y = b_0 + b_1 x_1 + b_2 x_2 \tag{4.17}$$

This is multiple regression analysis, which is an essential part of the response surface
methodology and model-dependent optimization techniques described in later chapters.

Regression of (4.17) involves three dimensions. Therefore, for visual representation,
a three-dimensional diagram is required. Many computer packages for experimental
design have the facility to produce these, which show a regression plane rather than
a regression line.

An illustration of the use of multiple regression analysis can be obtained from
the work of Evans et al.[4] The objective of their experiment was to investigate the
carminative (or flatulence-relieving) properties of a series of 26 volatile compounds.
All compounds possessed a substituent group containing an oxygen atom linked to

hydrogen, an alkyl group or an alkoxy group. The hypothesis was that carminative action was dependent on two factors. The first was the bulkiness of the smaller group attached to oxygen, expressed by the van der Waal's volume (V_W, measured in nm^3). The second was the octanol–water partition coefficient of the compounds (P). The response (ID_{50}) was the concentration ($M \times 10^3$) needed to reduce the response to a standard dose of carbachol by 50%.

The general equation (4.17) can be expressed as (4.18)

$$\log\left(\frac{1}{ID_{50}}\right) = b_0 + b_v V_W + b_P \log P \tag{4.18}$$

The data for the 26 compounds are shown in Table 4.6, and regression yields (4.19)

TABLE 4.6
Substituent Groups with Their Molar Volumes, Partition Coefficients, and Carminative Activities of a Series of Volatile Compounds

Compound	Substituent Group	V_w (nm³) (x_1)	log P (x_2)	log (1/ID_{50}) ($M \times 10^3$) (y)
Isobutanol	H	0.22	0.74	0.77
n-Butyl acetate	CH₃C=O	3.64	1.74	1.36
1,2-Dihydroxybenzene	H	0.22	0.95	1.02
1,3-Dihydroxybenzene	H	0.22	0.79	1.05
1,4-Dihydroxybenzene	H	0.22	0.55	0.91
1-Cresol	H	0.22	1.95	1.64
2-Cresol	H	0.22	1.99	1.54
3-Cresol	H	0.22	1.93	1.54
Dibutyl ether	CH₃(CH₂)₃	6.51	3.06	1.23
Diethyl ether	CH₃CH₂	3.41	0.80	0.59
3,4-Dimethylphenol	H	0.22	2.42	1.91
Di-isopropyl ether	(CH₃)₂CH	4.97	1.63	0.71
Di-n-propyl ether	CH₃(CH₂)₂	4.97	3.03	1.00
Ethyl acetate	CH₃C=O	3.64	0.70	0.59
Ethylvinyl ether	CH₂=CH	3.01	1.04	1.21
Eugenol	H	0.22	2.99	2.43
1-Hexanol	H	0.22	2.03	1.47
Menthol	H	0.22	3.31	2.13
2-Methoxyphenol	H	0.22	1.90	1.26
4-Methoxyphenol	H	0.22	1.34	1.32
1-Pentanol	H	0.22	1.16	1.11
2-Phenoxyethanol	H	0.22	1.16	0.90
Isopropyl acetate	CH₃C=O	3.64	1.02	0.96
n-Propyl acetate	CH₃C=O	3.64	1.50	0.94
Salicylaldehyde	H	0.22	1.76	1.70
Thymol	H	0.22	3.30	2.66
Total		41.17	44.79	33.95
Mean		1.58	1.72	1.31
SD		2.01	0.86	0.53

$\Sigma(y^2) = 51.42$; $\Sigma(x_1^2) = 166.29$; $\Sigma(x_2^2) = 95.55$; $\Sigma(x_1y) = 41.75$; $\Sigma(x_2y) = 67.14$; $\Sigma(x_1x_2) = 73.65$

Source: Evans et al.[4]

$$\log\left(\frac{1}{ID_{50}}\right) = 0.670 - 0.132V_W + 0.490\log P \qquad (4.19)$$

As with linear regression, multiple regression analysis can be used to predict values of the dependent variables for given values of the independent variables. Thus, by substituting the appropriate values of V_W and $\log P$ into (4.19), predicted values of $\log(1/ID_{50})$ can be obtained. These are given in Table 4.7.

If multiple regression is carried out using Microsoft Excel, the output is as shown in Table 4.8.

TABLE 4.7
Measured Values of Carminative Activity of a Series of Volatile Compounds and Values Predicted from (4.19) and (4.22)

| | log (1/ID$_{50}$) (M × 10³) (y) | | |
| | | Predicted From | |
Compound	Measured	(4.19)	(4.22)
Isobutanol	0.77	1.00	0.88
n-Butyl acetate	1.36	1.04	1.09
1,2-Dihydroxybenzene	1.02	1.11	1.01
1,3-Dihydroxybenzene	1.05	1.03	0.91
1,4-Dihydroxybenzene	0.91	0.91	0.77
1-Cresol	1.64	1.60	1.60
2-Cresol	1.54	1.62	1.62
3-Cresol	1.54	1.59	1.58
Dibutyl ether	1.23	1.31	1.02
Diethyl ether	0.59	0.61	0.76
3,4-Dimethylphenol	1.91	1.83	1.87
Di-isopropyl ether	0.71	0.81	0.92
Di-n-propyl ether	1.00	1.50	1.31
Ethyl acetate	0.59	0.53	0.71
Ethylvinyl ether	1.21	0.78	0.88
Eugenol	2.43	2.12	2.21
1-Hexanol	1.47	1.64	1.64
Menthol	2.13	2.26	2.40
2-Methoxyphenol	1.26	1.57	1.57
4-Methoxyphenol	1.32	1.30	1.24
1-Pentanol	1.11	1.21	1.13
2-Phenoxyethanol	0.90	1.21	1.13
Isopropyl acetate	0.96	0.69	0.83
n-Propyl acetate	0.94	0.93	1.01
Salicylaldehyde	1.70	1.50	1.48
Thymol	2.66	2.26	2.39

Source: Evans et al.[4]

TABLE 4.8
Microsoft Excel Output after Multiple Regression of the Data Given in Table 4.6, Fitting the Data to an Equation of the Form of (4.17)

A	B	C
1	0.490 (coefficient b_2)	−0.132 (coefficient b_1)	0.670 (coefficient b_0)
2	0.055 (standard error of b_2)	0.023 (standard error of b_1)	0.109 (standard error of b_0)
3	0.822 (coefficient of determination)	0.234 (standard error of Y)	
4	53.18 (F)	23 (degrees of freedom)	
5	5.827 (regression sum of squares)	1.260 (residual sum of squares)	

As before, additional parameters are required to support the validity of the regression equation. The more important of these follow.

4.4.1 CORRELATION COEFFICIENTS

There are four correlation coefficients associated with (4.17). Three are linear correlation coefficients, one for each combination of two variables, that is, r_{x_1y}, r_{x_2y}, and $r_{x_1x_2}$, and these are calculated according to (4.6). The other is the coefficient of multiple regression, r_{y,x_1x_2}, which applies to the complete equation. It can be calculated from (4.20)

$$r_{y,x_1x_2} = \sqrt{\frac{\left(r_{x_1y}\right)^2 + \left(r_{x_2y}\right)^2 - 2r_{x_1y}r_{x_2y}r_{x_1x_2}}{1 - \left(r_{x_1x_2}\right)^2}} \qquad (4.20)$$

If the correlation coefficient between x_1 and x_2 is significant, then the so-called independent variables are not truly independent but are related. In this situation, one should consider ignoring either x_1 or x_2 and working with a simpler relationship such as (4.1).

The linear correlation coefficients for the data in Table 4.6 are $r_{x_1y} = 0.758$, $r_{x_2y} = -0.449$, and $r_{x_1x_2} = 0.063$. The low value of $r_{x_1x_2}$ suggests that the two independent variables are unlikely to be related.

Hence, the coefficient of multiple regression is obtained by substitution into (4.20)

$$r_{y,x_1x_2} = \sqrt{\frac{0.758^2 + 0.449^2 - 2[0.758 \times (-0.449) \times 0.063]}{1 - 0.063^2}} = 0.907$$

What constitutes a satisfactory correlation coefficient is dependent on the purpose for which it is to be used and on the nature of the raw data. For a given number of sets of data, the more variables that are considered, the better the coefficient of multiple regression will appear to be. For example, if there are two variables and two pairs of results, linear regression analysis will inevitably give a correlation coefficient of 1.000, even if the numbers have been chosen at random, because the best fit of any two points is a straight line. Similarly, if we are trying to relate five systems, and data on five variables are available, then the more variables that are drawn into the correlation, the better will be the coefficient of multiple regression, until when all five variables are considered, the coefficient will equal 1.000. The resulting equation will be a perfect fit, and can be used as a model for that specific data, but is of little value in predicting new data. A useful point is that five regression points are the minimum necessary for each independent variable in an equation used for prediction purposes. Thus, if there are two independent variables, there should be at least ten sets of data points. This obviously has major implications for the design of an experiment.

4.4.2 STANDARD ERROR OF THE COEFFICIENTS AND THE INTERCEPT

These are calculated in a similar manner to that shown for linear regression and are displayed in the same way. However, the calculation is more protracted.

4.4.3 *F* VALUE

This has the same meaning as in linear regression analysis and is displayed in the same manner. An additional degree of freedom is subtracted for each additional variable. Thus, if n is the number of sets of data and m the number of variables in the regression equation, the F value will be displayed as $F_{(m-1),(n-m)}$.

4.5 INTERACTION BETWEEN INDEPENDENT VARIABLES

An interaction may occur between independent variables, in that the response to a change in one independent variable is governed by the value of the second independent variable. If this is so, an interaction term is introduced into (4.17), giving (4.21)

$$y = b_0 + b_1 x_1 + b_2 x_2 + b_{12} x_1 x_2 \tag{4.21}$$

The simplest way of solving an equation of this type is to calculate the interaction terms $x_1 x_2$ beforehand and introduce them as another independent variable with the coefficient b_{12}. Applying the data in Table 4.6 yields the regression equation (4.22), with a correlation coefficient of 0.907. The full Microsoft Excel output is shown in Table 4.9.

$$\log\left(\frac{1}{ID_{50}}\right) = 0.449 + 0.0015 V_W + 0.602 \log P - 0.064 V_W \log P \tag{4.22}$$

The predicted values of $\log(1/ID_{50})$ using this equation are given in Table 4.7.

TABLE 4.9
Microsoft Excel Output after Multiple Regression of the Data Given in Table 4.6, Fitting the Data to an Equation of the Form of (4.21)

	A	B	C	D
1	−0.064 (coefficient b_{12})	0.602 (coefficient b_2)	0.0015 (coefficient b_1)	0.449 (coefficient b_0)
2	0.022 (standard error of b_{12})	0.062 (standard error of b_2)	0.0507 (standard error of b_1)	0.122 (standard error of b_0)
3	0.871 (coefficient of determination)	0.204 (standard error of Y)		
4	49.36 (F)	22 (degrees of freedom)		
5	6.170 (regression sum of squares)	0.917 (residual sum of squares)		

If there are three independent variables x_1, x_2, and x_3, then the regression equation becomes (4.23)

$$y = b_0 + b_1 x_1 + b_2 x_2 + b_3 x_3 + b_{12} x_1 x_2 + b_{13} x_1 x_3 + b_{23} x_2 x_3 + b_{123} x_1 x_2 x_3 \qquad (4.23)$$

There are three two-way interaction terms, and the final term in the equation represents a three-way interaction. Because four dimensions would be required, such relationships cannot be portrayed graphically.

A further possibility is to combine nonlinear regression and multiple regression in the form of an equation such as (4.24)

$$y = b_0 + b_1 x_1 + b_2 x_2 + b_{11} x_1^2 + b_{22} x_2^2 + b_{12} x_1 x_2 \qquad (4.24)$$

4.6   STEPWISE REGRESSION

The introduction of additional variables, all or some of which may be raised to powers of 2 or more, together with interaction terms, will lead to long and complicated regression equations. Note that not all terms in the equation might be important. It is therefore necessary to find out which combination of independent variables shows the best relationship to the response. Stepwise regression is one way of achieving this.

The dependent variable is first regressed with each independent variable in turn, and the independent variable that alone gives the highest value of r^2 is selected. In the second step, the dependent variable is regressed against the selected independent variable in conjunction with each of the rejected variables in turn, giving a series of three-variable equations. The combination giving the highest value of r^2 is then selected and the process repeated with each of the remaining independent variables, plus the two selected variables. The process is continued within the confines of the amount of experimental data available, and the value of each additional predictor can be judged from the improvement in r^2.

From the data in Table 4.6, regression of $\log(1/ID_{50})$ against $\log P$, V_W, and the interaction term $\log P \cdot V_W$ gives r^2 values of 57.5%, 20.1%, and 5.5%, respectively. Thus, the highest value of r^2 is obtained by regressing the dependent variable against $\log P$. The next stage is to regress $\log(1/ID_{50})$ against the combinations of $\log P$ and V_W and $\log P$ and $\log P \cdot V_W$, which gives r^2 values of 82.2% and 87.1%, respectively. Lastly, regression of the dependent variable against all three independent variables, including the interaction term, gives a value of r^2 of 87.1%.

Stepwise regression techniques have been used to investigate several problems in pharmaceutical fields. For example, Wehrle et al.[5] have used this technique to study the relative importance of factors involved in wet granulation in a high shear mixer. Bohidar et al.[6] used stepwise regression to identify the relative importance of five factors in a tablet formulation — diluent, compression pressure, disintegrant concentration, amount of granulating fluid, and lubricant concentration — in relation to ten tablet properties. Tattawasart and Armstrong[7] identified the important factors in producing plugs of lactose for hard-shell capsule fills by using a similar technique.

4.7 RANK CORRELATION

In all the correlation techniques described so far, the variables have been measured on a continuous scale, that is, they have truly numerical values. The difference between a value of 1 and a value of 2 is the same as, say, the difference between 5 and 6. In Chapter 3, nonparametric tests were discussed after data had been placed in rank order. Here, there is no certainty that the difference between the first and second ranks is the same as that between the fifth and sixth. In such cases, the rank correlation coefficient, often called the Spearman coefficient of rank correlation after its originator, can be used.

This is best illustrated by an example. A pharmaceutical company makes eight different oral liquids, designated here A to H. Each formulation contains sucrose. A tasting panel evaluates all eight for palatability and draws up an order of preference, with the most palatable scoring 1 and the least 8.

Subsequently, sucrose is replaced by a synthetic sweetening agent, the medicines are assessed for palatability as before, and a second rank order is constructed.

The information is shown in Table 4.10.

The Spearman rank correlation coefficient (r_s) is given by (4.25)

$$r_s = 1 - \frac{6 \sum d^2}{n(n^2 - 1)} \tag{4.25}$$

where
  d = the difference in ranking awarded to each medicine
  n = the number of medicines, in this case 8.

Substitution into (4.25) gives

$$r_s = 1 - \frac{6 \times 52}{8(64-1)} = +0.381$$

As in many statistical tests, the calculated value of r_s is compared with a tabulated value. Values of Spearman's rank correlation coefficient are given in Table 4.11.

The tabulated value of r_s for a sample size of 8 at the 5% level is 0.714; therefore, it can be concluded that, at this probability level, there is no significant difference between the rankings.

If both rankings had agreed exactly, Σd^2 would equal 0 and r_s would equal +1. If one ranking had been the reverse of the other, then r_s would have a value of −1. The convention of omitting the plus or minus sign before the value of the correlation coefficient should not therefore be applied in this case.

The table of values of the Spearman coefficient goes up to a value of $n = 10$. For larger sample sizes, r_s is approximately related to Student's t, with $n-2$ degrees of freedom, by (4.26).

$$t = r_s \sqrt{\frac{n-2}{1-r_s^2}} \qquad (4.26)$$

Hence, if n exceeds 10, r_s is converted to t using (4.26) and the result compared with the appropriate tabulated value of t.

TABLE 4.10
Ranking of Eight Oral Liquids for Palatability

Product		A	B	C	D	E	F	G	H	
Sweetening agent	Sucrose	1	2	5	8	4	6	7	3	
	Synthetic	2	4	1	6	5	3	8	7	
	d	−1	−2	4	2	−1	3	−1	−4	$\Sigma d = 0$
	d^2	1	4	16	4	1	9	1	16	$\Sigma d^2 = 52$

TABLE 4.11
Values of Spearman's Rank Correlation Coefficient that Differ Significantly from Zero at the 5% and 1% Levels, Using a Two-Tail Test

Sample Size(%)	5	6	7	8	9	10
0.05	1.000	0.886	0.750	0.714	0.683	0.648
0.01	—	1.000	0.893	0.857	0.833	0.794

The value of r_s in Table 4.11 is 1.000 for a sample size of 5 at the 5% level of significance. Therefore, 5 is the minimum sample size that can be used for this test, and if this minimum is used, even identical ranking will not give a calculated value of r_s in excess of the tabulated value. This has implications for the design of the experiment.

4.8   COMMENTS ON THE CORRELATION COEFFICIENT

The computed value of the correlation coefficient is an indication of the goodness of fit to the type of equation that is assumed. Thus, if a linear equation is assumed, and (4.3) gives a value of r which is close to 0, then there is almost no linear correlation between the variables. It does not mean that there is no correlation at all. If a higher-order equation had been assumed, fit might have been improved and a higher value of r obtained.

It is essential to distinguish between correlation and causality. If the correlation coefficient of y on x has a high value (i.e., near -1 or $+1$), this does not necessarily mean that a change in x causes a proportionate change in y. Simply because two variables are correlated does not necessarily mean that one is the cause and the other is the effect.

The ideal experimental situation is where the value of one or more factors is deliberately changed by a known amount, everything else is kept constant, and the response is measured. In such cases, there will be little doubt as to which is cause and which is effect. However, even in a laboratory situation, this ideal is often difficult to achieve, and outside the laboratory, keeping all other factors constant is virtually impossible, particularly when time forms an independent variable. There is apparently a strong positive correlation, from the 1930s to 1940s, between sales of radio sets and admissions to mental hospitals. Which is cause and which is effect? Does owning a radio lead to insanity, or do only insane people buy radios? Or is this a spurious correlation? The two "responses" are in no way related but are merely both changing in the same direction at the same time.

A more subtle spurious correlation is that of purchases of cars and television sets in the U.K. in the second half of the 20th century. This is not an example of cause and effect but rather that both of them are the effects of the same cause, namely, that they reflect a rise in prosperity that enabled the two purchases to be made. Just because a correlation exists between two variables does not necessarily mean that they are related.

FURTHER READING

Robinson, E. A., *Least Squares Regression Analysis in Terms of Linear Algebra*, Goose Pond Press, Houston, 1981.

Bou-Chaira, N. A., Pinto, T. D., and O'Hara, M. T., Evaluation of preservative systems in a sunscreen formula by the linear regression method, *J. Cosmet. Sci.*, 54, 1, 2003.

Gohel, M. C. and Jogani, P. D., Exploration of melt granulation technique for the development of coprocessed directly compressible adjuvant containing lactose and microcrystalline cellulose, *Pharm. Dev. Technol.*, 8, 175, 2003.

Hariharan, M., Wheatley, T. A., and Price, J. C., Controlled release target matrices from carrageenans: compression and dissolution studies, *Pharm. Dev. Technol.*, 2, 383, 1997.

Lee, K. J. et al., Evaluation of critical formulation factors in the development of a rapidly dispersing captopril oral dosage form, *Drug Dev. Ind. Pharm.*, 29, 967, 2003.

Merrku, P. and Yliruusi, J., Use of 3^3 factorial design and multilinear stepwise regression analysis in studying the fluidised bed granulation process, *Eur. J. Pharm. Biopharm.*, 39, 75, 1993.

Nazzal, S. et al., Optimisation of a self-nanoemulsified tablet dosage form of ubiquinone using response surface methodology: effect of formulation ingredients, *Int. J. Pharm.*, 240, 103, 2002.

Pinto, J. F., Podczeck, F., and Newton, J. M., Investigation of tablets prepared from pellets produced by extrusion and spheronisation. II. Modelling the properties of the tablets produced using linear regression, *Int. J. Pharm.*, 152, 7, 1997.

Rambali, B., Baert, L., and Massart, D. L., Scaling up of the fluidised bed granulation process, *Int. J. Pharm.*, 252, 197, 2003.

Rambali, B. et al., Using deepest regression method for optimisation of fluidised bed granulation on semi full scale, *Int. J. Pharm.*, 258, 85, 2003.

Stephens, D. et al., A statistical experimental approach to cosolvent formulation of a water insoluble drug, *Drug Dev. Ind. Pharm.*, 25, 961, 1999.

REFERENCES

1. Clarke, G. M. and Cooke, D., *A Basic Course in Statistics*, 4th ed., Arnold, London, 1998.

2. Edwards, A. L., *An Introduction to Linear Regression and Correlation*, 2nd ed., Freeman, San Francisco, 1984.

3. Gebre-Mariam, T. et al., The use of electron spin resonance to measure microviscosity, *J. Pharm. Pharmacol.*, 43, 510, 1991.

4. Evans, B. K., James, K. C., and Luscombe, D. K., Quantitative structure–activity relationships and carminative activity, *J. Pharm. Sci.*, 67, 277, 1978.

5. Wehrle, P. et al., Response surface methodology: interesting statistical tool for process optimisation and validation: example of wet granulation in a high shear mixer, *Drug Dev. Ind. Pharm.*, 19, 1637, 1993.

6. Bohidar, N. R., Restaino, F. A., and Schwartz, J. B., Selecting key pharmaceutical formulation factors by regression analysis. *Drug Dev. Ind. Pharm.*, 5, 175, 1979.

7. Tattawasart, A. and Armstrong, N. A., The formation of lactose plugs for hard shell capsule fills, *Pharm. Dev. Technol.*, 2, 335, 1997.

5 Multivariate Methods

5.1 INTRODUCTION

Regression analysis, as discussed in the previous chapter, looks for relationships between a dependent variable (the response) and one or more independent variables (the factors). These are called univariate methods. Multivariate methods, on the other hand, look for relationships between random variables, considering them collectively. Multivariate methods initially consider each relationship to be equally important and then go on to assess which variables are related and which are not. As with so many other statistical techniques, the approach to multivariate methods has been revolutionized by the ready availability of computing power.

Multivariate analysis usually involves transformation of raw data into some form of matrix in which relationships are more easily identified. Hence, use is made of matrix algebra, and a brief introduction to this is given in Appendix 2. Some types of matrices relevant to experimental interpretation are discussed in this chapter.

Manly[1] provides a good general introduction to multivariate methods. Their application to pharmaceutical systems has been reviewed by Lindberg and Lundstedt.[2]

5.2 MULTIVARIATE DISTANCES

Many multivariate methods consist of measuring "distances," either between observations or populations. The simplest case is where there are n individuals, each of which has values for p variables $X_1, X_2, \ldots, X_p$.

5.2.1 DISTANCE MATRICES

In these, distances between individual observations are determined. This is illustrated in the following example.

A frequently encountered task is to find alternative sources of excipients for pharmaceutical formulations. Materials from all sources must meet analytical specifications, but it is obviously desirable that the ingredients from the alternative sources resemble the original material as closely as possible. Multivariate analysis can help in finding solution to such problems by constructing a distance matrix.

A topical semisolid contains olive oil, and the oil currently used is from source A. The specification for the olive oil prescribes an acceptable range for two properties, namely, acid value and iodine value. The relevant data for source A are shown in Table 5.1. Two alternative sources of oil become available (B and C), and their acid and iodine values are also shown in Table 5.1. The question to be resolved is whether the oil from source B or from source C most closely resembles oil from

TABLE 5.1
Analytical Data for Three Samples of Olive Oil

Sample	Acid Value	Iodine Value
A	0.1	79
B	0.5	82
C	0.2	88
Mean	0.267	83.000
Standard deviation	0.208	4.583

source A. The iodine value of B is nearer to that of A, but the acid value of C is closer to that of A.

The technique is to compare the three samples in two-dimensional space. The three pairs of data points are shown in Figure 5.1, with both axes drawn on the same scale.

The distance between samples (the Euclidean distance) is calculated by Pythagoras' theorem. Thus, the distance between A and B is given by (5.1)

$$(AB)^2 = (\text{acid value difference})^2 + (\text{iodine value difference})^2 \qquad (5.1)$$

Substitution from Table 5.1 gives

$$AB^2 = (0.5 - 0.1)^2 + (82 - 79)^2 = 9.160$$

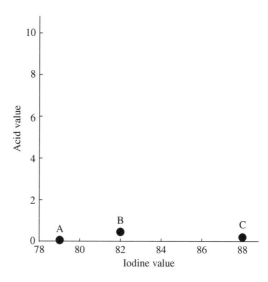

FIGURE 5.1 Graphical representation of analytical data from three samples of olive oil, using raw data.

Therefore, AB=3.027.

The corresponding value for AC=9.000.

One problem should be apparent at this stage, that the two properties have very different orders of magnitude. The acid values of all three samples are less than 1, whereas all iodine values are about 80. Therefore, even minor differences in iodine value will, on a numerical basis, greatly outweigh any differences in acid value. The distance between A and B is almost totally controlled by the iodine values, and the influence of change in the acid values is negligible. Even if the acid value of B is doubled to 1.0, the distance AB is only changed to 3.132.

Therefore, the elements of the matrix must be standardized so that all are of equal importance. This is done by subtracting the mean of the column from each element in it and dividing the result by the standard deviation of that column. Thus, the standardized value of the acid value of sample A is

$$\frac{0.1-0.267}{0.208} = -0.803$$

The standardized results are shown in Table 5.2 and Figure 5.2. A characteristic of the standardized results is that the sum of the elements in each column is zero and the standard deviations are unity.

Substituting the standardized values from Table 5.2 into (5.1) gives

$$(AB)^2=(-0.803-1.120)^2+[-0.873-(-0.218)]^2=4.127$$

Therefore, AB=2.031.

Similarly, the distance from A to C is given by

$$(AC)^2=[-0.803-(-0.322)]^2+(-0.873-1.090)^2=4.085$$

Therefore, AC=2.021.

Hence, oil from source C more closely resembles that from source A than does that from source B.

The same technique can be used for any number of properties and any number of samples. For three properties, graphical representation is by a three-dimensional

TABLE 5.2
Standardized Values of the Analytical Data for Three Samples of Olive Oil

Sample	Acid Value	Iodine Value
A	−0.803	−0.873
B	1.120	−0.218
C	−0.322	1.090

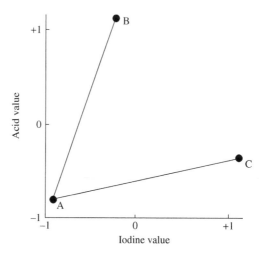

FIGURE 5.2 Graphical representation of analytical data from three samples of olive oil, using standardized data.

diagram, but this is impossible for more than three properties. Despite this, calculation of the distance between samples can be achieved using the Pythagoras theorem as described above. Thus, for example, imagine there are five sources of olive oil (A, B, C, D, and E) and five different properties (acid value, iodine value, refractive index, saponification value, and weight per milliliter). The individual values are shown in Table 5.3 and their standardized values in Table 5.4. Note that the standardized values of acid and iodine values for samples A, B, and C differ from those shown in Table 5.2 because the means and standard deviations are now different.

The distance between A and B is given by

$$\sqrt{(-1.266-1.266)^2+\left[-1.309-(-0.416)\right]^2+(-0.526-0.351)^2+(-0.816-0.000)^2+\left[-0.956-(-0.956)\right]^2}$$

$$=2.939$$

TABLE 5.3
Analytical Data for Five Samples of Olive Oil

Sample	Acid Value	Iodine Value	Refractive Index	Saponification Value	Weight per ml (g)
A	0.1	79	1.469	192	0.911
B	0.5	82	1.470	193	0.911
C	0.2	88	1.471	192	0.912
D	0.4	83	1.468	195	0.913
E	0.3	85	1.470	193	0.912
Mean	0.300	83.400	1.470	193.000	0.912
Standard deviation	0.158	3.362	0.001	1.225	0.001

TABLE 5.4
Standardized Analytical Data for Five Samples of Olive Oil

Sample	Acid Value	Iodine Value	Refractive Index	Saponification Value	Weight per ml
A	−1.266	−1.309	−0.526	−0.816	−0.956
B	1.266	−0.416	0.351	0.000	−0.956
C	0.633	1.368	1.228	−0.816	0.239
D	−0.633	−0.119	−1.403	1.633	1.434
E	0.000	0.476	0.351	0.000	0.239

All computed distances are shown in the distance matrix Table 5.5. Therefore, oil from source E most closely resembles that from source A, taking all five properties into consideration.

The numbers below the leading diagonal are a mirror image of those above it. For this reason, this half of the matrix is usually omitted.

Distance matrices as described above are useful for establishing how closely one material resembles another. The various properties are independent, and there is no suggestion that they are in any way related to or dependent upon each other.

5.3 COVARIANCE MATRICES

A common problem is to establish whether one property (termed the dependent variable or response) is governed by the magnitude of one or more of the other properties (the independent variables or the factors). This has already been covered in the chapter on regression and is covered again in chapters on response surface methodology and modeling, but covariance and correlation matrices offer an alternative approach.

The work of James et al.[3] offers a useful example. These workers attempted to establish a structure–activity relationship involving the androgenic activities of five esters of testosterone. Their hypothesis was that activity might depend on one or more of three properties of the esters that were related to their structures. These were the catalytic rate constant (k_c), the partition coefficient (R_m), and the bulkiness of the ester group (E_s). The experimental data, together with full definitions of the

TABLE 5.5
Distance Matrix for Five Samples of Olive Oil

Sample	A	B	C	D	E
A	—	2.939	3.908	3.781	2.766
B	2.939	—	2.539	3.891	1.955
C	3.908	2.539	—	4.261	1.621
D	3.781	3.891	4.261	—	2.815
E	2.766	1.955	1.621	2.815	—

terms used, are given in Table 5.6. Conversion of these data to a covariance matrix reveals more information on their interdependence.

The variance of a column of elements is the square of the standard deviation of those elements. The variance (V) is equal to the sum of the squares of the differences between each element in the column and their mean, divided by the number of elements minus 1, as expressed in (5.2). Sometimes, in multivariate analysis, n is used rather than $n - 1$. However, it is more convenient to use $n - 1$ to bring results into line with normal statistical practice. It makes no difference in the long run which denominator is used, provided it is used consistently.

$$V = \frac{\sum (x - x_\mathrm{m})^2}{n - 1} \tag{5.2}$$

where
  x_m = the mean
  x = individual values
  n = the number of elements.

Variance is more easily computed from (5.3)

$$V = \frac{\sum (x^2) - \sum (x)^2 / n}{n - 1} \tag{5.3}$$

TABLE 5.6
Androgenic Activities and Quantitative Structure–Activity Relationship (QSAR) Parameters of Some Testosterone Esters

Ester	Log Overall Androgenic Response (log OAR)	Log Catalytic Constant (log k_c)	R_m	E_s
Formate	1.63	1.27	0.58	0.00
Acetate	2.04	1.48	0.46	−1.24
Propionate	2.70	2.00	0.11	−1.58
Butyrate	2.96	2.09	−0.09	−1.60
Valerate	2.84	2.06	−0.26	−1.63
Mean	2.434	1.780	0.160	−1.210
Standard deviation	0.573	0.378	0.356	0.695

Note: Overall androgenic response represents the area under the curve obtained when the weights of prostate gland plus seminal vesicles of castrated rats were plotted against time since dosing. Catalytic constant (k_c) is the rate constant for the *in vitro* hydrolysis of the esters with standardized liver homogenate. R_m is a chromatographic parameter derived from the R_f value and logarithmically related to partition coefficient.[4] E_s is a parameter related to the bulkiness of the ester group.[5]

Source: James et al.[3]

where

$\Sigma(x^2)$ = the sum of the squares of all the elements in that column

$\Sigma(x)^2/n$ = the square of the sum of all the elements divided by the number of elements.

Thus, the variance of the first column of elements in Table 5.6 is calculated as follows

$$\Sigma(x^2) = (1.63^2 + 2.04^2 + 2.70^2 + 2.96^2 + 2.84^2) = 30.9357$$

$$\frac{\sum (x)^2}{n} = \frac{(1.63 + 2.04 + 2.70 + 2.96 + 2.84)^2}{5} = 29.6218$$

Thus, the variance = $(30.9357 - 29.6218)/4 = 0.328$.

The covariance (c_{xy}) between a column of elements (x) and another column of elements (y) is given by (5.4) but can be more easily calculated from (5.5)

$$c_{xy} = \frac{\sum (y - y_m)(x - x_m)}{n - 1} \tag{5.4}$$

$$c_{xy} = \frac{\sum (xy) - \left[\sum (x) \sum (y)/n\right]}{n - 1} \tag{5.5}$$

$\Sigma(xy)$ is the sum of the products of x and y and $\Sigma(x)\Sigma(y)$ the product of the sums of x and y. n now represents the number of pairs of elements, x and y.

From the second and third columns of Table 5.6

$$\Sigma(xy) = [(1.63 \times 1.27) + (2.04 \times 1.48) + (2.70 \times 2.00) + (2.96 \times 2.09) + (2.84 \times 2.06)] = 22.5261$$

$$\Sigma(x) = 12.17 \text{ and } \Sigma(y) = 8.90$$

Therefore, $\Sigma(x)\Sigma(y)/n = 21.6626$.

Therefore, the covariance between the elements in columns 1 and 2 (V_{12})

$$= \frac{22.5261 - 21.6626}{5 - 1} = 0.216$$

Covariance values between all the columns are shown in Table 5.7.

TABLE 5.7
Covariance Matrix Derived from Data in Table 5.6

	log OAR	log k_c	R_m	E_s
log OAR	0.328	0.216	−0.193	−0.359
log k_c		0.143	−0.128	−0.232
R_m			0.127	0.198
E_s				0.483

If every value in column 1 of Table 5.6 had been identical, then $\Sigma(x-x_m)$ would be zero and (5.4) reduces to zero. A covariance of zero therefore indicates no relationship between the two columns, and following this, the greater the covariance, the more likely there is to be a relationship.

The results shown in Table 5.7 suggest that the dependent variable or response (log OAR) is more dependent on the steric factor E_s with a covariance of −0.359 than on either log k_c, with a value of 0.216, or R_m, with a value of −0.193. However, there are no criteria with respect to covariances to suggest which values are encouraging and which are not.

An important feature of covariance matrices is that they are always square matrices, even when the matrices from which they have been derived are not. The importance of this is that several parameters associated with multivariate analysis, for example, determinants, can only be calculated for square matrices. This is the situation with Table 5.6, in which there are five rows and four columns, but determinants can be calculated after conversion to the covariance matrix shown in Table 5.7.

5.4 CORRELATION MATRICES

Table 5.8 shows the standardized form of the data given in Table 5.6.

As the data in Table 5.8 have been standardized, the sum of each column is zero, and the standard deviation (and hence the variance) of each column is unity. Substitution of values of columns 2 and 3 of Table 5.8 into (5.5) yields (5.6)

TABLE 5.8
Standardized Values of Androgenic Activities and Quantitative Structure–Activity Relationship (QSAR) Parameters of Some Testosterone Esters

Ester	log OAR	log k_c	R_m	E_s
Formate	−1.403	−1.349	1.180	1.741
Acetate	−0.688	−0.794	0.843	−0.043
Propionate	0.464	0.582	−0.140	−0.532
Butyrate	0.918	0.820	−0.702	−0.561
Valerate	0.708	0.741	−1.180	−0.604

Source: James et al.[3]

TABLE 5.9
Correlation Matrix Derived from Data in Table 5.6

	log OAR	log k_c	R_m	E_s
log OAR	1.000	0.997	−0.945	−0.901
log k_c	0.997	1.000	−0.948	−0.882
R_m	−0.945	−0.948	1.000	0.800
E_s	−0.901	−0.882	0.800	1.000

$$c_{xy} = \frac{(-1.403 \times -1.349) + \cdots + (0.708 \times 0.741)}{5-1} = 0.997 \tag{5.6}$$

The covariance matrix of the standardized values given in Table 5.8 is shown in Table 5.9 and displayed in cross-reference form in the same way as Table 5.7. An identical result is obtained by linearly regressing each column of elements in turn with the other columns and displaying the correlation coefficients. Each element in a correlation matrix is equal to the correlation coefficient between the row and the column in which it lies. Therefore, Table 5.9 is usually described as a correlation matrix.

Table 5.9 reveals that the logarithm of the catalytic constant is rectilinearly related to the logarithm of the overall androgenic response ($r=0.997$), but a relationship between log OAR and R_m ($r=-0.945$) is also indicated. However, the intersection of the R_m column with the log k_c row ($r=-0.948$) suggests that this may be explained by relationships between R_m and log k_c. Further tests would be necessary to resolve these issues and will be described in the book as they arise.

Note that all elements in the leading diagonal of Table 5.9 are equal to unity and that the elements in the top right-hand half of the matrix are reflected across the leading diagonal. For this reason, one half of the table is frequently omitted when this type of matrix is presented.

5.5 CLUSTER ANALYSIS

5.5.1 CARTESIAN PLOTS

Cluster analysis is a form of multivariate analysis that is used to solve the following problem. There are n objects, each of which is assessed by means of p variables. Can these objects be sorted into groups or clusters so that "similar" ones are in the same group? It is a useful technique for assessing preliminary data when quantitative results have not yet been established. However, because the data are not quantitative, regression methods cannot be used.

The work of McFarland and Gans[6] provides a good example. The objective of this work was to relate the monoamine oxidase-inhibiting (MAOI) properties to Hansch Π values,[7] which assess lipophilicity, and Taft substituent parameters (E_s),[5] which are a measure of the bulk of the substituting group in the molecule. McFarland and Gans[6] studied the activities of 20 compounds of similar structure, and so the two groups are "active" and "inactive." Seven compounds were found to be active. Table 5.10 summarizes the results.

Two one-dimensional plots can be prepared from the results, as shown in Figure 5.3. A one-dimensional plot is of course a straight line. Figure 5.3(a) is scaled in terms of Hansch Π values, and Figure 5.3(b) uses Taft substituent parameters. Scrutiny of the plots reveals that, with the exception of a few outliers, the inactive compounds are clustered toward the high Π values and the low E_s values, suggesting that biological activity is dependent on low lipid solubility and the absence of large substituent groups.

An alternative is to prepare a two-dimensional plot, as shown in Figure 5.4. The positions of the points confirm the dependence of monoamine oxidase inhibition on steric factors, and the fact that all the active compounds have low Π values confirms the importance of low lipid solubility. The procedures applied to these results are explained in more detail in McFarland and Gans's article, together with application to more complicated systems. It can be used equally well with quantitative data by choosing an activity threshold, below which the observation is considered to represent inactivity. The precise value of the threshold is not critical, because decisions are based on recognition of patterns, which allows latitude with respect to the level at which the threshold is pitched.

TABLE 5.10
Activities and Physicochemical Parameters of 20
Monoamine Oxidase Inhibitors[6]

Compound	Active (+) or Inactive (−)	Π	E_s
1	+	1.3	0.00
2	+	1.2	0.32
3	+	1.3	0.32
4	+	2.2	−0.07
5	+	1.7	0.00
6	+	1.0	0.00
7	+	0.8	0.32
8	−	1.7	0.00
9	−	1.7	−0.66
10	−	2.7	−0.66
11	−	4.2	−0.68
12	−	3.5	−0.68
13	−	1.0	−0.66
14	−	1.0	0.00
15	−	2.6	−1.08
16	−	2.6	−1.08
17	−	2.1	−1.08
18	−	0.8	0.32
19	−	1.4	−0.66
20	−	4.7	−0.68

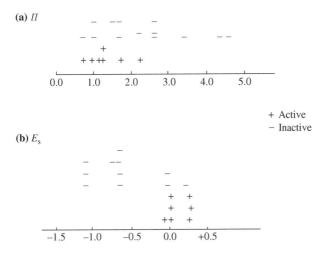

FIGURE 5.3 One-dimensional cluster plot of (a) monoamine oxidase-inhibiting (MAOI) activity against Π and (b) MAOI activity against E_s (data taken from Table 5.10).

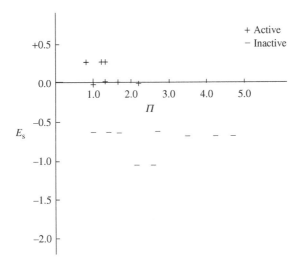

FIGURE 5.4 Two-dimensional cluster plot of monoamine oxidase-inhibiting (MAOI) activity against Π and E_s (data taken from Table 5.10).

5.5.2 DENDROGRAMS

In an example earlier in this chapter, five samples of olive oil were compared to determine which two samples were nearest to each other with respect to five analytical properties. Cluster analysis can be used as an extension of this, in which samples are classified into clusters of nearest neighbors. The procedure does not determine the number of groups, but given the number of groups required, it selects which samples go into which clusters. Considering the five samples of olive oil, the samples initially fall into five groups: A, B, C, D, and E. The distance matrix (Table 5.5) shows that C and E are closest in properties; therefore, if we wish to classify the data into four clusters, we would combine C and E, which are separated by only 1.621 units, to give (C, E), A, B, and D.

B and E are the next nearest to each other (1.955); therefore, for three clusters, the arrangement is (B, C, E), A, and D. For two clusters, it is (A, B, C, E) and D (2.766), and for one cluster, the next nearest neighbors are D and E, with a distance of separation of 2.815 units. This information can be plotted in the form of a dendrogram, in which the clusters are arranged along the abscissa and the distances between the clusters from the ordinate. The dendrogram can be plotted in terms of nearest neighbors, as in Figure 5.5(a), or of the furthest neighbor distance, as shown in Figure 5.5(b). The latter has the same overall shape as the nearest neighbor plot, but the heights of some of the blocks are greater. The distance between C and E is fixed at 1.621 units, and so this block has the same height in both plots, but the second cluster takes the greatest distance between a pair from B, C, and E, which is 2.539 units. Similarly, the furthest distances for A, B, C, E and A, B, C, D, E are 3.908 and 4.261, respectively.

Note that the samples are not arranged in alphabetical order in Figure 5.5. This is because the samples should be arranged in a manner in which the information is most easily understood. Thus, in the present situation, if A, B, C, D, and E were arranged in alphabetical order, it would be impossible to plot a dendrogram.

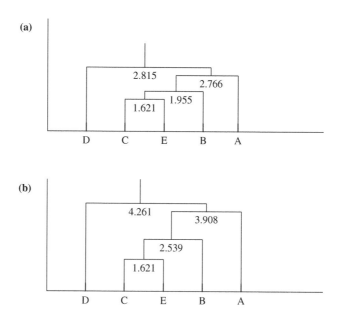

FIGURE 5.5 Dendrograms of analytical data from five samples of olive oil (data taken from Table 5.4): (a) nearest neighbor plot, (b) furthest neighbor plot.

5.6 DISCRIMINATION ANALYSIS

Discrimination analysis can be regarded very loosely as the reverse of cluster analysis. In cluster analysis, the data are processed as a whole, with the object of identifying groups of related results, whereas in discrimination analysis the data are initially divided into groups, according to a preconceived hypothesis, and the credibility of the classification then assessed. In the simplest situation, the hypothesis that a collection of values of one variable is divisible into two subgroups can be tested by plotting the variable on a scatter diagram, in the same way as was employed with clustering. The hypothesis can be tested by visual observation and is characterized by the points separating into two groups. A subsequent, more sophisticated treatment could use a test for significance, such as the Student's t test, from which the probability of there being two groups can be assessed.

Typical results with two variables are shown in Figure 5.6. The variables may be directly related, giving one or two straight lines [Figure 5.6(a) and (b)], or giving elliptical plots [Figure 5.6(c) and (d)]. The existence of one or two groups can be judged from the degree or absence of overlap of points, which have been tentatively allocated to different groups. Alternatively, the two variables may be independent, giving scattered plots [Figure 5.6(e) and (f)], but Figure 5.6(f) also provides a method of discriminating between two groups.

Scatter diagrams can therefore be used to establish the existence of two or more subgroups within the complete data set. They can also be used to allocate new results to their various sets by ascertaining where the results lie in the diagram. The process has the advantage that the number of individual results within each group need not be the same. However, it is essential that the variances within the groups be similar; otherwise, the outcome would be biased in favor of the variables with the greatest variance. For this reason, it is advisable to standardize the data before analysis is carried out.

When assignment to a group is ambiguous, it becomes necessary to calculate to which cluster an individual result belongs. Such calculations are dependent on the distance between the point and a position representative of the profile, usually the mean coordinates. Thus, in two dimensions, the distance (d) between a point (x, y) and the average value of cluster A (x_m, y_m) is given by (5.7). Similarly, the distance from the average of cluster B is given by (5.8), so that if $d_A < d_B$, the point belongs to cluster A and vice versa.

$$d_A = \sqrt{(x - x_m)^2 + (y - y_m)^2} \qquad (5.7)$$

$$d_B = \sqrt{(x - x_m)^2 + (y - y_m)^2} \qquad (5.8)$$

The method can be extended to any number of dimensions.

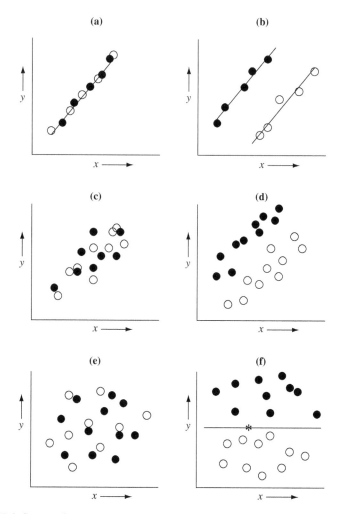

FIGURE 5.6 Scatter diagrams: (a) x related to y, one group; (b) x related to y, two groups; (c) x correlated with y, one group; (d) x correlated with y, two groups; (e) no correlation, one group; (f) no correlation, two groups.

Discriminant analysis can be demonstrated using data published by Rushton.[8] Rushton was interested in the structure of hair, and the objective of that work was to investigate whether male and female hair differed according to three criteria. These were:

1. The total number of hairs per square centimeter of scalp.
2. The number of hairs per square centimeter with a diameter greater than 40 μm and a length greater than 30 mm. These values were chosen because they are the lower limits of the dimensions of fibers that contribute to the aesthetic quality of the hair.
3. The number of actively growing hairs per square centimeter of scalp, expressed as a percentage of the total number of hairs.

Rushton's results are shown in Table 5.11.

Scores in each column are standardized by subtracting the mean for males plus females and dividing by the corresponding standard deviation. Thus, for example, the standardized total number of hairs on the first male in Table 5.11 is given by (5.9)

$$\text{standardized result} = \frac{177 - 210.2}{45.0} = -0.738 \tag{5.9}$$

Standardized means are given in Table 5.11. The root mean square distances between each standardized result and the male and female standardized means are then calculated. For example, for the first male, comparison with the average male gives

TABLE 5.11
Parameters for Male and Female Hair[8]

	Total Hairs		Hairs >40 μm in Diameter and >30 mm in Length		Actively Growing Hair	
	Number	Standardized	Number	Standardized	%	Standardized
Males						
1	177	−0.738	154	−0.600	61.9	−1.052
2	175	−0.782	159	−0.480	48.6	−2.209
3	180	−0.671	167	−0.287	59.8	−1.235
4	231	0.462	219	0.966	73.4	−0.052
5	222	0.262	193	0.340	75.2	0.104
6	218	0.173	174	−0.118	61.5	−1.087
7	240	0.662	208	0.701	79.5	0.478
8	239	0.640	205	0.629	72.8	−0.104
9	276	1.462	234	1.328	67.4	−0.574
10	269	1.307	241	1.496	89.3	1.330
Mean	222.7	0.278	195.4	0.398	68.9	−0.440
Standard deviation	36.3		31.1		11.6	
Females						
1	125	−1.893	108	−1.708	86.8	1.113
2	154	−1.249	138	−0.986	80.1	0.530
3	156	−1.204	137	−1.010	75.2	0.104
4	172	−0.848	113	−1.588	75.1	0.096
5	236	0.573	223	1.063	88.0	1.217
6	259	1.084	158	−0.504	73.1	−0.078
7	245	0.773	210	0.749	89.9	1.383
Mean	192.4	−0.395	155.3	−0.569	81.2	0.624
Standard deviation	53.0		45.2		6.99	
Statistical Parameters for Male Plus Female						
Mean	210.2		178.9		74.0	
Standard deviation	45.0		41.5		11.5	

TABLE 5.12
Euclidean Distances between Male and Female Hair Parameters

	Male		Female	
	Versus Male	**Versus Female**	**Versus Male**	**Versus Female**
1	1.550	1.711	3.408	1.942
2	2.241	2.859	2.278	0.947
3	1.414	1.899	2.115	1.051
4	0.712	1.888	2.346	1.231
5	0.547	1.241	1.992	1.809
6	0.834	1.861	1.647	1.262
7	1.040	1.665	1.921	1.923
8	0.545	1.747		
9	1.511	2.918		
10	2.323	2.773		

$$d_A = \sqrt{(-0.738 - 0.278)^2 + (-0.600 - 0.398)^2 + [-1.052 - (-0.440)]^2} = 1.550$$

Similarly, comparison with the average female gives

$$d_B = \sqrt{[-0.738 - (-0.395)]^2 + [-0.600 - (-0.569)]^2 + (-1.052 - 0.624)^2} = 1.711$$

1.550 is less than 1.711, so Male 1 is assigned to the male group. The complete results are presented in Table 5.12.

In all cases, the distances between male subjects and the average male are less than the distances from the average female. For similar reasons, the females fall into a separate group. There is one exception, in that Female 7 gives virtually the same Euclidean distance from both male and female means. This must be considered to be an interface result, as exemplified by the point marked with an asterisk in Figure 5.6(f). It may be concluded that, according to the three parameters quoted, male hair is different from female hair, and therefore, discrimination into two groups is justified.

5.7 PRINCIPAL COMPONENTS ANALYSIS

Principal components analysis is one of the simplest multivariate methods and has been widely used for interpreting experimental data in pharmaceutical fields. Its object is to reduce the number of variables of possible importance in characterizing an array of numbers. The data are transformed into a small number of linear combinations of the original variables, called principal components. If there are n variables, $X_1, X_2, X_3, \ldots, X_n$, there will be n possible principal components, $Z_1, Z_2, Z_3, \ldots, Z_n$, that are uncorrelated. Lack of correlation is important, because it means that the variables represent different aspects of the data. Correlation, on the other hand, means that the same aspect is measured in more than one way. There would thus be redundancy in the data, and the experimental design could be simplified. The indices are ordered, so that Z_1 displays the highest amount of variation, Z_2 the second highest, and so on, and when combined they will explain the whole of the variance of the data.

The aim of a principal components analysis is to establish whether or not any of the indices has such a low value that its effect is negligible. This means that variation in the data can be described by a smaller number of indices whose effect is not negligible. If all the original indices are unrelated, then principal components analysis will achieve nothing, in that the number of variables will remain unchanged.

The quantitative structure–activity relationship (QSAR) study of androgenic activities of testosterone esters described earlier[3] will be used as an example. The overall androgenic response described in Table 5.6 may be divided into two parts: the maximum weight achieved by the organ (the maximum biological response BR_{max}) and the time taken to achieve that weight (time of maximum response T_{max}). The logarithms of these, together with parameters previously described in Table 5.6, are given in Table 5.13.

The first step in carrying out a principal components analysis on these data is to convert them to a standardized form, as described earlier. These are given in Table 5.14.

A correlation matrix is then constructed from the standardized data, also by using methods described earlier in this chapter. The correlation matrix is given in Table 5.15.

TABLE 5.13
Androgenic Activities and Quantitative Structure–Activity
Relationship (QSAR) Parameters of Some Testosterone Esters[3]

Ester	log T_{max}	log BR_{max}	log k_c	R_m	E_s
Formate	0.30	1.60	1.27	0.58	0.00
Acetate	0.48	1.73	1.48	0.46	−1.24
Propionate	0.78	2.10	2.00	0.11	−1.58
Butyrate	0.78	2.31	2.09	−0.09	−1.60
Valerate	0.90	2.02	2.06	−0.26	−1.63
Mean	0.648	1.952	1.780	0.160	−1.210
Standard deviation	0.249	0.286	0.378	0.356	0.695

TABLE 5.14
Standardized Values of Androgenic Activities and Quantitative
Structure–Activity Relationship (QSAR) Parameters Shown in Table 5.13

Ester	log T_{max}	log BR_{max}	log k_c	R_m	E_s
Formate	−1.398	−1.231	−1.349	1.180	1.741
Acetate	−0.675	−0.776	−0.794	0.843	−0.043
Propionate	0.530	0.517	0.582	−0.140	−0.533
Butyrate	0.530	1.252	0.820	−0.702	−0.561
Valerate	1.020	0.238	0.741	−1.180	−0.604

TABLE 5.15
Correlation Matrix of Standardized Androgenic Activity Data from Table 5.14

	log T_{max}	log BR_{max}	log k_c	R_m	E_s
log T_{max}	1.000	0.856	0.980	0.967	0.900
log BR_{max}	0.856	1.000	0.945	0.835	0.808
log k_c	0.980	0.945	1.000	0.948	0.883
R_m	0.967	0.835	0.948	1.000	0.800
E_s	0.900	0.808	0.883	0.800	1.000

The next stage is to calculate eigenvalues and eigenvectors for the data given in Table 5.14. These calculations are too protracted to carry out manually but are readily processed by computer. This yields Table 5.16.

Some general rules with respect to a principal components display are:

1. Each row of eigenvectors is called a principal component, and the sum of the squares of the terms in each principal component is always equal to unity. Thus, for the first principal component of Table 5.16, the sum of squares is

$$0.46^2 + 0.43^2 + 0.47^2 + (-0.45)^2 + (-0.43)^2 = 1.00$$

2. The elements running from the top left-hand corner to the bottom right-hand corner of a matrix form the leading diagonal. The sum of the elements in the leading diagonal of the original matrix is always equal to the sum of the eigenvalues. Thus, in Table 5.15

sum of the elements in the leading diagonal $= (1 + 1 + 1 + 1 + 1) = 5$

and in Table 5.16

sum of the values $= (4.57 + 0.21 + 0.20 + 0.02 + 0.00) = 5$

TABLE 5.16
Eigenvectors and Eigenvalues for Androgenic Activity Data

Principal Component (Z_n)	Eigenvalues	Eigenvectors (F_n)				
		log T_{max}	log BR_{max}	log k_c	R_m	E_s
1	4.57	0.46	0.43	0.47	−0.45	−0.43
2	0.21	−0.03	−0.21	−0.14	0.44	−0.86
3	0.20	0.37	−0.80	−0.10	−0.46	−0.04
4	0.02	−0.54	0.18	−0.47	−0.62	−0.27
5	0.00	−0.60	−0.31	0.73	−0.09	−0.07

3. The eigenvalues are all positive. This is always so when they are derived from a correlation matrix.
4. Each eigenvalue expresses the fraction of the variance of the elements of the matrix that is explained by the component. Thus, the first principal component explains

$$\frac{4.57 \times 100}{5} = 91.4\%$$

of the variance.
5. The sum of the squares of the eigenvectors in each principal component is called the communality of the row. As stated above, this should equal 1. The importance of an eigenvector in a row can therefore be assessed by calculating the communality without that vector and by noting how far it deviates from unity. Thus, the communality of the first principal component without the first term is

$$1.00 - 0.46^2 = 0.79$$

which indicates that this term is important. In contrast, the communality of the third principal component without the last term is

$$(1.00 - 0.04^2) = 0.998$$

indicating that this term is not of importance to its principal component. Scrutiny of Table 5.16 therefore tells us:

1. 91.4% of the variance of the data is explained by an expression involving all five variables, that is,

$$0.46 \log T_{max} + 0.43 \log BR_{max} + 0.47 \log k_c - 0.45 R_m - 0.43 E_s$$

2. $(0.21 \times 100)/5 = 4.2\%$ can be explained by the second principal component. Communalities without the first and third terms are

$$[1.00 - (-0.03)^2] = 1.00$$

and

$$[1.00 - (-0.14)^2] = 0.98$$

indicating that 4.2% of the variance is explained by an expression involving $\log BR_{max}$, R_m, and E_s.
3. 4.0% of the variance is explained by an expression involving the two biological responses, $\log T_{max}$ and $\log BR_{max}$, together with R_m.

4. The remaining principal components explain only 0.4% and 0.0% of the variance. The fourth and fifth principal components can therefore be ignored. A zero eigenvalue also suggests a rectilinear relationship within the variables. Thus, two of the variables may be measuring the same thing, or one variable plus another variable may add up to 100%, that is, definition of one automatically defines the other. There would thus be scope for simplifying the design of future experiments.

5. The signs of the eigenvectors in Table 5.16 indicate that 99.6% of the variance involves inverse relationships between the biological parameters and partition, and 95.4% involves inverse relationships between duration of biological action and the size of the ester group.

The information therefore suggests that the net biological response is a combination of three mechanisms, with one mechanism predominant. However, the amount of raw data presented does not allow much confidence in this conclusion, because a comparatively simple system has been used to illustrate a statistical technique to make the procedures easier to follow. As stated before, five extra points are the minimum necessary for each additional variable.

Useful information can sometimes be obtained by plotting the eigenvectors for one principal component against the corresponding eigenvectors of another component. The first two principal components are usually chosen for this purpose.

A much larger series of experiments was carried out by Benkerrour et al.,[9] who used principal components analysis to study granule and tablet properties. Using three diluents (lactose, tricalcium phosphate, and a 1:1 mixture of the two) and two varieties of guar gun as binding agents, each at concentrations of 0.5%, 1.0%, and 1.5%, they prepared 18 granule formulations. For each formulation, they measured seven properties of the granules. The granules were then compressed into tablets, and five tablet properties were then determined. There were thus 12 items of information for each of the 18 formulations. They constructed a 12×12 correlation matrix of the granule and tablet properties and found that some of the properties were very highly correlated, for example, tablet pore volume and surface area had a correlation coefficient of 0.994. The former was also highly correlated with granule pore volume, applied pressure, and tablet hardness, and thus they were in effect measuring the same thing. Benkerrour et al. were also able to show that the nature of the diluent was much more important than the type or concentration of binder.

Another large study was performed by Hogan et al.[10] These workers used a design of 33 experiments to study the relationship between formulation factors and the filling into and active ingredient release from hard-shell gelatin capsules. Nine formulation properties, for example, drug solubility, disintegrant type, and concentration, were considered, and for each formulation, five properties related to filling performance (e.g., bulk density, variation of fill weight) were assessed, together with five related to drug release, such as disintegration time and characteristics of the dissolution curve. An interesting feature of this work was that the data were examined by both parametric and nonparametric methods. Most of the data were numerical, but some, for example, filler type, were not. For parametric analysis, the nonnumerical data were allocated a dummy numerical value, whereas for nonparametric analysis, the numerical data were transformed into ordinal forms by putting them into rank order. Some of the responses were found to be highly correlated with others and hence could be excluded from future experiments.

Other pharmaceutical applications of principal components analysis are given in the bibliography of this chapter.

5.8   FACTOR ANALYSIS

This branch of multivariate analysis is closely related to principal components analysis. The basic premise is that it may be possible to describe a set of n variables, $X_1, X_2, \ldots, X_n$, in terms of a smaller number of indices, which will help establish a relationship between the variables. Unlike principal components analysis which is not based on any specific statistical model, factor analysis is based on a model first reported by Spearman.[11] In his seminal paper, he studied correlations between test scores and found that many correlations could be accounted for by a simple model. Expressing his data as a correlation matrix, he noted that if the diagonals of the matrix are ignored, any two rows are often proportional.

Factor analysis can be applied to many types of experimental data but has most frequently been used to assess examination results, as in the following example.

Twenty students each take examinations in six subjects, designated A to F. The results are given in Table 5.17.

The normal procedure is to arrange the subjects in columns, and average the numbers along each row, to give the overall performances of the candidates. By placing these scores in numerical order, a rank order of achievement is obtained. In a similar way, the standards in each subject can be compared by calculating the means of the columns.

A candidate's score in a particular subject can be resolved into two factors: the candidate's ability and the degree of difficulty of the subject, which can be expressed mathematically in the form of (5.10)

$$x = aF \qquad (5.10)$$

where
   $x =$ the score obtained by a given candidate in the given subject
   $F =$ a constant, specific to the subject and independent of the candidate
   $a =$ a constant, specific to the candidate.

The concept can be extended to the full diet of subjects by using (5.11) for the first candidate and by similar equations for each of the remaining candidates.

$$X_1 = a_{1A}F_A + a_{1B}F_B + a_{1C}F_C + a_{1D}F_D + a_{1E}F_E + a_{1F}F_F \qquad (5.11)$$

These equations, embracing 20 dependent variables and 120 independent variables, are difficult to handle statistically. It would be useful to simplify the picture by principal components analysis, but this is not possible, because the independent variables form a 6×20 matrix, and principal components analysis can be applied only to square matrices.

The problem can be overcome by using the corresponding correlation matrix, as given in Table 5.18, which is derived from the standardized values of the data presented in Table 5.17.

TABLE 5.17
The Scores for 20 Students in 6 Examinations

Candidate Number	Subject Score (%)						Aggregate Score	Class Position
	A	B	C	D	E	F		
1	30	51	44	38	35	37	235	20
2	48	43	61	52	58	50	312	12
3	52	54	72	68	59	51	356	6
4	41	46	56	65	56	24	288	15
5	52	62	65	46	61	57	343	9
6	56	67	72	73	51	49	368	4
7	51	43	58	42	62	57	313	11
8	41	40	51	53	54	57	296	14
9	48	68	59	58	56	55	344	8
10	56	87	70	73	65	66	417	1
11	44	63	57	43	52	46	305	13
12	51	69	62	71	60	44	357	5
13	58	69	75	63	68	67	400	2
14	48	56	45	47	51	39	286	16
15	57	71	70	71	65	63	397	3
16	53	58	55	56	52	40	314	10
17	42	48	54	42	54	37	277	17
18	35	43	50	39	50	41	238	19
19	36	59	47	50	39	42	273	18
20	45	65	66	66	55	48	345	7
Mean	47.2	58.1	59.5	55.8	55.2	48.5		
Standard deviation	7.8	12.2	9.4	12.2	8.1	11.0		

TABLE 5.18
Correlation Matrix of Standardized Examination Scores from Table 5.17

	A	B	C	D	E	F
A	1.000	0.582	0.814	0.642	0.904	0.643
B	0.582	1.000	0.579	0.626	0.351	0.487
C	0.814	0.579	1.000	0.740	0.758	0.645
D	0.642	0.626	0.740	1.000	0.503	0.323
E	0.904	0.351	0.758	0.503	1.000	0.648
F	0.643	0.487	0.645	0.323	0.648	1.000

If the leading diagonal results are ignored, the ratios of the numbers in any pair of rows are approximately constant. For example, with rows A and C in Table 5.18 and if columns A and C are ignored, the ratios A/C are all about equal to unity. Thus,

$$\frac{0.582}{0.579} = 1.0, \quad \frac{0.642}{0.740} = 0.9, \quad \frac{0.643}{0.645} = 1.0$$

From results like these, Spearman speculated that examination results could be simplified by ignoring some of the subjects, using (5.12)

$$x = aF + e \qquad (5.12)$$

where
x = the score obtained by a given candidate in a given subject
F = a constant, the factor value, specific to the subject and independent of the candidate
e = related to both subject and candidate.

If a subject, or a column, is ignored, e represents the error involved in making the approximation. From (5.13) it can be assessed that

$$e = (1 - \text{communality of remaining terms}) \qquad (5.13)$$

The operation of factor analysis depends on the transposition of matrices. A matrix is said to be transposed when the rows are interchanged with the columns, as shown, for example, in (5.14)

$$\begin{bmatrix} a & b \\ c & d \end{bmatrix} \text{ is the transpose of } \begin{bmatrix} a & c \\ b & d \end{bmatrix} \qquad (5.14)$$

This procedure is permissible with principal components. Therefore, for two candidates 1 and 2, and two subjects A and B, the principal components Z_1 and Z_2 are given by (5.15) and (5.16), respectively

$$Z_1 = b_{1A}x_{1A} + b_{2A}x_{2A} \qquad (5.15)$$
$$Z_2 = b_{2A}x_{2A} + b_{2B}x_{2B} \qquad (5.16)$$

These can be transposed to (5.17) and (5.18).

$$X_1 = b_{A1}z_{A1} + b_{B1}z_{B1} \qquad (5.17)$$
$$X_2 = b_{A2}z_{A2} + b_{B2}z_{B2} \qquad (5.18)$$

where

X = a candidate's overall performance
x = that candidate's performance in one subject
Z = principal component for all subjects
z = principal component for one subject.

Transposition of the equations requires scaling the principal components so as to have unit variances. This is done by dividing the values of Z by their standard deviations, which are equal to the square roots of the corresponding eigenvalues, to give the factors (F), that is,

$$F = \frac{Z}{\sqrt{\lambda}} \tag{5.19}$$

Eigenvalues and eigenvectors are shown in Table 5.19.

Substitution of the eigenvectors for b, and $F\sqrt{\lambda}$ for Z in the coefficients shown in (5.17) and (5.18) then gives (5.20) to (5.25).

For example, the first term on the right-hand side of (5.20) is equal to $-0.45 \times F_A\sqrt{4.08} = -0.91F_A$, the second term equal to $-0.11 \times F_B\sqrt{0.82} = -0.10F_B$, and the third term equal to $-0.46 \times F_C\sqrt{0.56} = +0.10F_C$. Similarly, the first term on the right-hand side of (5.21) is $-0.36 \times F_A\sqrt{4.08} = -0.73F_A$.

$$X_1 = -0.91F_A - 0.10F_B + 0.10F_C + \cdots \tag{5.20}$$
$$X_2 = -0.73F_A + 0.49F_B - 0.44F_C + \cdots \tag{5.21}$$
$$X_3 = -0.93F_A - 0.00F_B + 0.15F_C + \cdots \tag{5.22}$$
$$X_4 = -0.79F_A + 0.49F_B + 0.28F_C + \cdots \tag{5.23}$$
$$X_5 = -0.83F_A - 0.40F_B + 0.24F_C + \cdots \tag{5.24}$$
$$X_6 = -0.75F_A - 0.41F_B - 0.44F_C + \cdots \tag{5.25}$$

Only the first three components need to be considered, because the eigenvalues indicate that these alone will explain over 90% of the variation, that is,

TABLE 5.19
Eigenvectors and Eigenvalues of Standardized Examination Scores from Table 5.17

Eigenvalues	Eigenvectors					
4.08	−0.45	−0.36	−0.46	−0.39	−0.41	−0.37
0.82	−0.11	0.54	0.00	0.54	−0.44	−0.45
0.56	0.14	−0.59	0.20	0.38	0.32	−0.59
0.25	−0.52	−0.37	0.36	0.40	−0.30	0.46
0.15	0.55	−0.18	0.44	−0.27	−0.63	−0.06
0.14	−0.44	0.25	0.66	0.42	0.21	−0.30

TABLE 5.20
Spearman Rank Order Correlation Matrix of Examination
Scores from Table 5.17

	A	B	C	D	E	F
A	1.000	0.631	0.811	0.636	0.693	0.655
B	0.631	1.000	0.603	0.639	0.394	0.406
C	0.811	0.603	1.000	0.732	0.735	0.694
D	0.636	0.639	0.732	1.000	0.467	0.397
E	0.693	0.394	0.735	0.467	1.000	0.734
F	0.655	0.406	0.694	0.397	0.734	1.000

$$\frac{(4.08 + 0.82 + 0.56) \times 100}{4.08 + 0.82 + 0.56 + 0.25 + 0.15 + 0.14} = 91.0\%$$

In equations (5.20) to (5.22), apart from the first term on the right-hand side, only one factor loading (0.49) is near 0.5, and none exceeds it. The conclusion therefore is that the overall abilities of most of the candidates can be estimated from their performances in one subject.

Problems of this type could also be examined using the rank order correlation coefficient described in Chapter 4. If the original marks in the six subjects are arranged into a rank order, the Spearman rank order correlation coefficient can be calculated for each pair of results. These are shown in Table 5.20. Results for subjects A and C are highly correlated with those of the other subjects.

FURTHER READING

Chatfield, C. and Collins, A. J., *Introduction to Multivariate Analysis*, Chapman & Hall, London, 1989.

Everitt, B., Landau, S., and Leese, M., *Cluster Analysis*, 4th ed., Arnold, London, 2001.

Kendall, M., *Multivariate Analysis*, 2nd ed., Griffin, High Wycombe, 1980.

Romesburg, H. C., *Cluster Analysis for Researchers*, Krieger, Malabar, 1984.

Adams, E. et al., Evaluation of dissolution profiles using principal components analysis, *Int. J. Pharm.*, 212, 41, 2001.

Baines, E., Factor analysis in the evaluation of cosmetic products, *J. Soc. Cosmet. Chem.*, 29, 369, 1978.

Bjerknes, K. et al., Evaluation of different formulation studies on air-filled polymeric microcapsules by multivariate analysis, *Int. J. Pharm.*, 257, 1, 2003.

Bohidar, N. R. and Bohidar, N. R., Multivariate analyses of a production formulation optimization experiment, *Drug Dev. Ind. Pharm.*, 20, 2165, 1994.

Dias, V. H. and Pinto, J. F., Identification of the most relevant factors that affect and reflect the quality of granules by application of canonical and cluster analysis, *J. Pharm. Sci.*, 91, 273, 2002.

Dyrstad, K., Veggeland, J., and Thomassen, C., Multivariate method to predict the water vapor diffusion rate through polypropylene packaging, *Int. J. Pharm.*, 188, 767, 1999.

Frutos, G., Frutos, P., and Alonso, M. A., Application of the statistical multivariate analysis to the release characterisation of matrix tablets, *Drug Dev. Ind. Pharm.*, 20, 2685, 1994.

Gabriellson, J. et al., Multivariate methods in the development of a new tablet formulation, *Drug Dev. Ind. Pharm.*, 29, 1053, 2003.

Hardy, I. J. and Cook, W. G., Predictive and correlative techniques for the design, optimisation and manufacture of solid dosage forms, *J. Pharm. Pharmacol.*, 55, 3, 2003.

Harris, A. J. et al., Assessment of auxiliary detergents in shampoo mixtures, *Cosmet. Perfum.*, 90, 23, 1975.

Horhota, S. T. and Aitken, C. L., Multivariate cluster analysis of pharmaceutical data using Andrews plots, *J. Pharm. Sci.*, 80, 85, 1991.

Jorgensen, A. et al., Hydrate formation during wet granulation studied by spectroscopic methods and multivariate analysis, *Pharm. Res.*, 19, 1285, 2002.

Magnusson, B. M., Pugh, W. J., and Roberts, M. S., Simple rules defining the potential of compounds for transdermal delivery or toxicity, *Pharm. Res.*, 21, 1047, 2004.

Onuki, Y., Morishita, M., and Takayama, K., Formulation optimisation of water-in-oil-water multiple emulsion for intestinal insulin delivery, *J. Control. Release*, 97, 91, 2004.

Persson, B. et al., Multivariate parameter evaluation of pharmaceutically important cellulose esters, *J. Pharm. Sci.*, 88, 767, 1999.

Rogers, L. J. and Adams, M. J., Factor analysis for pharmaceutical solutions, *Pharm. Sci.*, 3, 333, 1997.

Romoran, K. et al., The influence of formulation variables on *in vitro* transfection efficiency and physicochemical properties of chitosan-based polyplexes, *Int. J. Pharm.*, 261, 115, 2003.

Rotthauser, B., Kraus, G., and Schmidt, P. C., Comparison of lubricants for effervescent tablets by principal components analysis, *Pharm. Ind.*, 60, 541, 1998.

Sande, S. A. and Dyrstad, K., A formulation strategy for multivariate kinetic responses, *Drug Dev. Ind. Pharm.*, 28, 583, 2002.

Westerhuis, J. A., Coenengracht, P. M., and Lerk, C. F., Multivariate modelling of the tablet manufacturing process with wet granulation and in-process control, *Int. J. Pharm.*, 156, 109, 1997.

REFERENCES

1. Manly, B. F., *Multivariate Statistical Methods – A Primer*, 2nd ed., Chapman & Hall, London, 1994.
2. Lindberg, N.-O. and Lundstedt, T., Application of multivariate analysis in pharmaceutical development work, *Drug Dev. Ind. Pharm.*, 21, 987, 1995.
3. James, K. C., Nicholls, P. J., and Richards, G. T., Correlation of androgenic activities of the lower testosterone esters in rat with R_m values and hydrolysis rates, *Eur. J. Med. Chem.*, 10, 55, 1975.
4. Bate-Smith, E. C. and Westall, R. G., Chromatographic behaviour and chemical structure. 1. Some naturally occurring phenolic substances, *Biochim. Biophys. Acta*, 4, 427, 1950.
5. Newman, M. S., Ed., *Steric Effects in Organic Chemistry*, Wiley, New York, 1956.
6. McFarland, J. W. and Gans, D. J., The significance of clusters in the graphical display of structure–activity relationships, *J. Med. Chem.*, 29, 505, 1986.

7. Iwasa, J., Fujita, T., and Hansch, C., Substituent constants for aliphatic functions obtained from partition coefficients, *J. Med. Chem.*, 8, 150, 1965.
8. Rushton, D. H., Chemical and Morphological Properties of Scalp Hair, Ph.D. thesis, University of Wales, 1988.
9. Benkerrour, L. et al., Granule and tablet formulae study by principal components analysis, *Int. J. Pharm.*, 19, 27, 1984.
10. Hogan, J. et al., Investigations into the relationship between drug properties, filling and release of drugs from hard gelatin capsules using multivariate statistical analysis, *Pharm. Res.*, 13, 944, 1996.
11. Spearman, C., "General intelligence," objectively determined and measured, *Am. J. Psychol.*, 15, 201, 1904.

6 Factorial Design of Experiments

6.1 INTRODUCTION

A classical approach to experimentation is to investigate the effects of one experimental variable while keeping all others constant. A well-known example would be the investigation of the relationship between the volume of a gas and pressure, keeping the temperature constant. This approach is valid provided that, as in this case, the underlying laws relating cause and effect are known with some certainty. However, in many cases, such knowledge is not available and it is not known, out of the many variables that might affect the outcome of an experiment, which will prove the most important and hence justify more extensive study.

Furthermore, it is possible that variables may interact with each other. Thus, the magnitude of the effect caused by altering one factor will depend on the magnitude of one or more of the other factors. An experimental design that investigates the effect of one factor while keeping all other factors at a constant level is unlikely to disclose the presence of such interactions.

An agricultural rather than a pharmaceutical example highlights the problem. Imagine that a study is to be carried out to compare the milk yields of Jersey cows and Highland cattle. If the test took place in an English meadow, the Jersey cows would be expected to have the highest yield. If, however, the site of the experiment were to be changed to a Scottish moor, the reverse result would be obtained, as the Jersey cows would probably not survive the harsher climate. Thus, the yield is dependent on both the breed of cow and the environment.

Factorial design, a technique introduced by Fisher[1] in 1926, is a system of experimental design that is intended to avoid such difficulties. It provides a means whereby the factors that may have an influence on a reaction or a process can be evaluated simultaneously and their relative importance assessed. It is thus a means of separating those factors that are important from those that are not. The technique can be applied to many pharmaceutical problems, and it forms the basis for many tests that seek to find an optimum solution.

The basis of the process is to elucidate the effects of many factors simultaneously, to assess their relative importance, and to determine whether the factors interact. All possible combinations of factors and levels are investigated, and thus all main effects and all interactions can be evaluated.

There are three decisions that have to be taken at the outset.

1. *The factors to be studied*: Factors can be quantitative, that is, they have numerical values, or they can be qualitative. The latter will often have

names rather than numbers, such as Method I, Site B, or Present or Absent. They will be chosen in accordance with the objectives of the experiment. Though the stated objectives of the experiment may define two or more factors, there may be other factors that would influence the outcome of the experiment, and these must be kept constant. A preliminary scanning program might be necessary to establish the relative importance of the factors.

2. *The levels of the factors*: This is often a difficult decision, in which the experience of the researcher plays an important part. A commonly used starting point is to select the 25th and 75th percentile levels of the range of possible values of the factor, though this might not be practicable.

3. *The response to be measured*: This is usually defined in the experimental objectives. The response must be capable of being expressed numerically. Adjectival descriptions (big, bigger, and biggest) or ordinal numerals (designating the biggest response as 1, the next biggest as 2, and so on) are not permissible.

6.2 TWO-FACTOR, TWO-LEVEL FACTORIAL DESIGNS

The simplest factorial design is one in which two factors are studied at two levels: low and high. The design consists of four experiments. These are:

Experiment A: Both factors are at their lower levels.
Experiment B: The first factor is at its higher level and the second at its lower.
Experiment C: The first factor is at its lower level and the second factor at its higher.
Experiment D: Both factors are at their higher levels.

Ideally, all four experiments are carried out simultaneously, the responses are measured, and the results are then assessed.

The procedure can be illustrated with the following example. Compound E is an ester and therefore would be expected to undergo hydrolysis when in aqueous solution. It is anticipated that the rate of hydrolysis will be influenced by temperature and the presence of a catalyst. Thus, the objective of the experiment is to ascertain the influence of two factors — temperature and catalyst concentration — on the rate of hydrolysis of Compound E. Two temperatures (20 °C and 40 °C) and two concentrations of catalyst (0 M and 0.1 M) are selected. The response that is measured is the loss of Compound E after a specified time. The four experiments are carried out simultaneously; the experimental conditions are shown in Table 6.1, which also shows the responses.

It is often helpful to envisage the experimental design as a diagram, which in this case is a square (Figure 6.1). Temperature forms the horizontal axis and the catalyst concentration the vertical axis.

The effect of the two factors can now be calculated. The effect of any given factor is the change in response produced by altering the level of that factor, averaged over the levels of all the other factors. Therefore, the effect of temperature is the mean of the results on the right-hand side of the square minus

TABLE 6.1
A Two-Factor, Two-Level Factorial Design to Study the Hydrolysis of Compound E

Experiment	Temperature (°C)	Catalyst Concentration (M)	Loss of E (%)
A	20	0	10
B	40	0	25
C	20	0.1	30
D	40	0.1	45

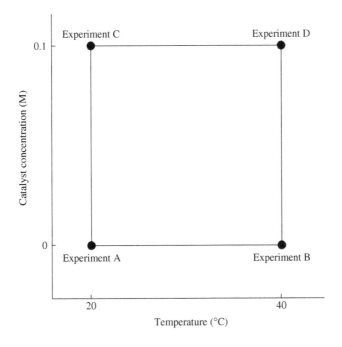

FIGURE 6.1 A two-factor, two-level experimental design.

the mean of those on the left-hand side. Similarly, the effect of the concentration of catalyst is the average of all results on the top of the square minus that of those results on the bottom.

Thus, the effect of temperature is given by (6.1)

$$\frac{1}{2}[(B+D)-(A+C)] \tag{6.1}$$

$$=\frac{1}{2}[(25+45)-(10+30)]=15$$

TABLE 6.2
A Two-Factor, Two-Level Factorial Design to Study the Hydrolysis of Compound E

Experiment	Temperature (°C)	Catalyst Concentration (M)	Loss of E (%)
A	20	0	10
B	40	0	25
C	20	0.1	70
D	40	0.1	95

Similarly, the effect of catalyst concentration is given by (6.2)

$$\frac{1}{2}\left[(C+D)-(A+B)\right] \tag{6.2}$$

$$=\frac{1}{2}\left[(30+45)-(10+25)\right]=20$$

Hence, in this example, both factors have an approximately equal effect and are therefore equally worthy of consideration.

However, consider an equally feasible alternative situation (Table 6.2). By the same method of calculation, the effect of temperature is 10 and the effect of the catalyst is 65. In this case, the catalyst concentration proves to have a more important effect, and attention should be focused on that.

The foregoing is a very straightforward example. However, the same principles can be used for much more complex systems.

6.2.1 TWO-FACTOR, TWO-LEVEL FACTORIAL DESIGNS WITH INTERACTION BETWEEN THE FACTORS

In the data presented in Table 6.1, an assumption has been made that the factors act independently to produce their effects. In many cases, this will be so, but in others, the level of one factor may govern the magnitude of the effect of another. This is termed factor interaction.

Interactions can often be detected graphically. In the data given in Table 6.1, raising the temperature causes an increased loss of Compound E, by 15% (25% − 10%). Similarly, the presence of a catalyst causes an increased loss of 20% (30% − 10%). When both factors are at a high level, the total increase in loss is 35% (45% − 10%), which is numerically equal to the total of the increased losses caused by the two factors considered separately. Thus, there is no interaction between the two factors. This situation is shown in Figure 6.2. A line is drawn joining the two results with catalyst concentration at a higher level (Experiments C and D) and another line joining the two experiments in which the same factor is at its lower level (Experiments A and B). If no interaction occurs, the lines will be parallel.

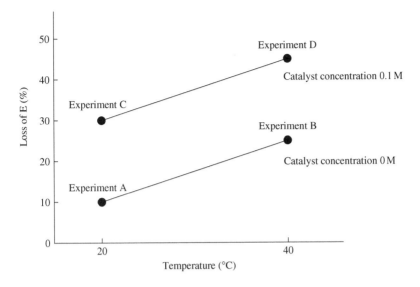

FIGURE 6.2 A two-factor, two-level experimental design with no interaction.

A quantitative estimation of factor interaction is made as follows. A further column is added to Table 6.1 and Table 6.2, "Factor interaction." For Experiment A, both the temperature and the catalyst concentration are at their lower levels, and hence "low×low" is entered in the "Factor interaction" column. Other rows in this column are completed in a similar manner. The results are shown in Table 6.3 and Table 6.4.

The interaction of temperature with catalyst concentration is given by (6.3)

$$\frac{1}{2}\left[(A+D)-(B+C)\right] \tag{6.3}$$

When the data from Table 6.3 are substituted into (6.3), the numerical value of the interaction is

$$\frac{1}{2}\left[(10+45)-(25+30)\right]=0$$

TABLE 6.3
A Two-Factor, Two-Level Factorial Design to Study the Hydrolysis of Compound E

Experiment	Temperature (°C)	Catalyst Concentration (M)	Factor Interaction	Loss of E (%)
A	20	0	Low × low	10
B	40	0	High × low	25
C	20	0.1	Low × high	30
D	40	0.1	High × high	45

TABLE 6.4
A Two-Factor, Two-Level Factorial Design to Study the Hydrolysis of Compound E

Experiment	Temperature (°C)	Catalyst Concentration (M)	Factor Interaction	Loss of E (%)
A	20	0	Low × low	10
B	40	0	High × low	25
C	20	0.1	Low × high	70
D	40	0.1	High × high	95

However, when data from Table 6.4 are used, the numerical value of the interaction is

$$\frac{1}{2}\left[(10+95)-(25+70)\right]=5$$

Thus, the interaction is not zero. If the data from Table 6.4 are plotted in an analogous manner to Figure 6.2, Figure 6.3 is obtained. The two lines are no longer parallel.

If the combined effect of the two factors is to produce a loss in Compound E greater than that produced by the factors individually, then the interaction is said to be synergistic. An interaction which produces a smaller loss is termed antagonistic.

Finding that an interaction has a significant effect can have a beneficial result. If, for example, an interaction occurs between the temperature of a reaction and the

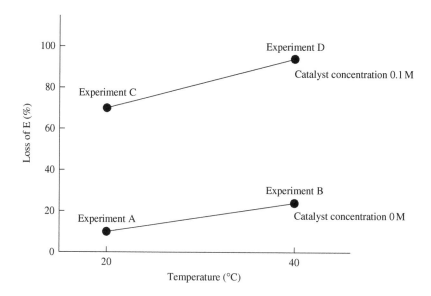

FIGURE 6.3 A two-factor, two-level experimental design with interaction.

quantity of catalyst present, then yield is highest either with a low temperature and a large amount of catalyst or with a high temperature and a small amount of catalyst. Hence, the most economical solution depending on the relative costs of energy and catalyst can be selected.

Caution must be exercised while interpreting Figure 6.2 and Figure 6.3. The pair of values of each factor has been joined together by a straight line. This line is purely illustrative, and no experimental evidence has been generated to support a rectilinear relationship. Further experiments at interpolated levels are needed to establish whether or not a factor and a response are related by a straight line.

6.3 NOTATION IN FACTORIALLY DESIGNED EXPERIMENTS

In the previous example, the four experiments were designated A, B, C, and D. These designations are simply labels and have no significance apart from that. Such a simple system is perfectly adequate for a two-factor, two-level design that is used solely to discover the relative importance of the two factors, that is, for screening purposes. However, if the design is to be used to derive response surfaces or for optimization purposes, a more elaborate system of notation is required. Several systems have been used, and they can often be the source of confusion.

Ideally, a system of notation should be selected that can be used for any experimental design, irrespective of the number of factors and the number of levels. Also, it should be capable of being used in conjunction with multiple regression techniques, which, as will be seen in later chapters, are the foundation of response-surface methods.

The notation that is used in the remainder of this chapter and in subsequent chapters is derived as follows. The factors are sequentially designated X_1, X_2, X_3, and so on. Thus, in the above example, temperature becomes X_1 and catalyst concentration X_2.

The actual values of the factors are now subject to a process known as "coding" and are expressed in terms of experimental units (e.u.). In Chapter 5, there is a worked example on distance matrices in which the properties of samples of olive oil were compared. Because the numerical values of the properties were so different (all iodine values were about 80, whereas all acid values were less than unity), it was necessary to standardize the data by subtracting the mean and dividing by the standard deviation. Coding is an analogous process that brings the values of all the factors into the same range. For a two-level experiment, the lower level is designated -1 and the upper level $+1$. Thus, for the factor temperature in the above example, 20 °C is designated -1 and 40 °C becomes $+1$. The range is 2 e.u. (-1 to $+1$), and as this equals 20 °C, 1 e.u. equals 10 °C. For the catalyst concentration, the lower concentration (0) is designated -1 and the higher (0.1 M) becomes $+1$. Therefore, 1 e.u. of catalyst concentration equals 0.05 M. Use of the range -1 to $+1$ permits the interpolation and definition of a central point (0, 0) in the design.

Table 6.3 and Figure 6.1 can now be reconstructed using this notation to form Table 6.5 and Figure 6.4, respectively. The values for the interaction term are obtained by multiplying together the values for the individual factors. Thus, for experiment (-1, -1), the value of the interaction is $-1 \times -1 = +1$.

TABLE 6.5
A Two-Factor, Two-Level Factorial Design to Study the Hydrolysis of Compound E, Using Coded Values of the Factors

Experiment	Factor X_1 (Temperature) (e.u.)	Factor X_2 (Catalyst Concentration) (e.u.)	Factor Interaction X_1X_2	Loss of E (%)
(−1, −1)	−1	−1	+1	10
(+1, −1)	+1	−1	−1	25
(−1, +1)	−1	+1	−1	30
(+1, +1)	+1	+1	+1	45

Note: e.u. = experimental unit; center point (0, 0): temperature = 30 °C, catalyst concentration = 0.05 M. 1 e.u. of temperature = 10 °C. 1 e.u. of catalyst concentration = 0.05 M.

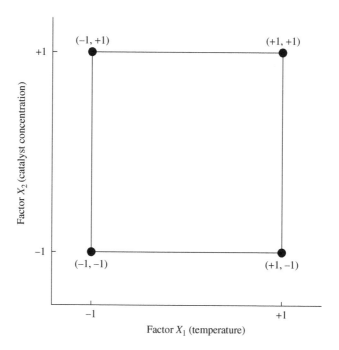

FIGURE 6.4 A two-factor, two-level experimental design, using coded values of the factors.

A common system of notation that can only be used in two-level studies is to designate the factors by uppercase letters, beginning with A. Any experiment in which a given factor is at a high level is designated by the corresponding lowercase letter. Thus, if only Factor A is at a high level, then that experiment is designated experiment "a," and if Factors A and B are both at high levels, the designation is "ab." The experiment in which all factors are at their lower level is denoted by (1). A further convention is to designate the lower level as "−" and the higher level as "+."

TABLE 6.6
A Two-Factor, Two-Level Factorial Design to Study the Hydrolysis of Compound E, Using an Alphabetical Notation

Experiment	Factor A (Temperature)	Factor B (Catalyst Concentration)	Factor Interaction AB	Loss of E (%)
(1)	−	−	+	10
a	+	−	−	25
b	−	+	−	30
ab	+	+	+	45

This is particularly useful when considering interactions between factors. Using this convention, Table 6.3 becomes Table 6.6.

There is also a convention in the order in which the experiments are written down in tabular form. A design is said to be in standard order when all the factors are at their lowest level for the first experiment, the first factor changes level every experiment, the second factor changes level every two experiments, the third factor every four experiments, and so on. Thus, the standard order for a two-factor, two-level design is $(-1, -1), (+1, -1), (-1, +1), (+1, +1)$. Hence, the first row in the table denotes the experiment in which both factors are at their lower levels, and the fourth is when both factors are at their higher levels. This convention is unimportant with two-factor experiments, but its usefulness will become more apparent with designs of greater complexity.

This convention relates to the order in which the experiments are written down. It is not the order in which they should be performed. Ideally, all experiments are carried out simultaneously, using identical apparatus and the same personnel. This is because there may be uncontrolled or even unknown factors that can affect the response. Simultaneous performance will ensure that all experiments are equally affected by these. If this is not possible, then the experiments should be carried out in random order.

6.4 FACTORIAL DESIGNS WITH THREE FACTORS AND TWO LEVELS

The previous discussion was limited to two experimental factors and a possible interaction between them. However, the principles of factorial design can be extended to situations in which many more factors can be examined.

Consider the situation in which three factors and their interactions are suspected of having an influence on the response. The procedures involved are best shown by means of a worked example.

Lactose is a commonly used diluent for tablet and capsule formulations. Though relatively inert, it can take part in the Maillard reaction to form small amounts of brown pigments, which in turn cause discoloration of the dosage form. Among the factors which may affect the rate of the reaction and hence the degree of discoloration are temperature and humidity. Because the Maillard reaction is base catalyzed and hence favored by alkaline conditions, the presence of a base would also be expected to increase discoloration.

Discoloration has been a particular problem with tablets containing spray-dried lactose, and Armstrong and Cartwright[2] examined varieties of lactose, both spray and conventionally dried, to determine their propensity to develop a brown color. The following factors and levels were selected:

Factor X_1: concentration of base (benzocaine) ($0\%=-1$; $5\%=+1$)
Factor X_2: temperature ($25\,°C=-1$; $40\,°C=+1$)
Factor X_3: humidity (50% RH$=-1$; 75% RH$=+1$)

The experiments were set up as shown in Table 6.7. Experiment ($+1$, $+1$, -1) is carried out in the presence of 5% benzocaine, the storage temperature is 40 °C, and the relative humidity is 50%. All experiments were carried out simultaneously, and after storage for 2 months in these conditions, tablet color was measured using a reflectance meter, pure white scoring zero. The greater the degree of discoloration, the higher the number. The results given in the table are those for lactose monohydrate.

Two points are worth mentioning at this stage. First of all, note that the tablet color must be expressed as a numerical value. Adjectival descriptions such as white and light brown cannot be used in designs of this type. Equally unacceptable are rank orders such as white$=1$ and the next lightest tablet$=2$. Next, note the standard order of the experiments in the table. The reason for adherence to this order will be apparent later.

Possible interactions must now be considered. In this case, there are three two-way interactions (Factor X_1 with Factor X_2, Factor X_1 with Factor X_3, and Factor X_2 with Factor X_3) and one three-way interaction (Factor X_1 with Factor X_2 and Factor X_3).

TABLE 6.7
Three-Factor, Two-Level Factorial Design to Investigate the Discoloration of Lactose Monohydrate Tablets

Experiment	Factor X_1 (Base Concentration) (%)	Factor X_2 (Temperature) (°C)	Factor X_3 (Humidity) (% RH)	Tablet Color
(−1, −1, −1)	0	25	50	1.6
(+1, −1, −1)	5	25	50	5.3
(−1, +1, −1)	0	40	50	3.4
(+1, +1, −1)	5	40	50	6.6
(−1, −1, +1)	0	25	75	2.6
(+1, −1, +1)	5	25	75	3.6
(−1, +1, +1)	0	40	75	3.0
(+1, +1, +1)	5	40	75	7.0

Note: Center point (0, 0, 0): base concentration $= 2.5\%$, temperature $= 32.5°C$, humidity $= 62.5\%$ RH. 1 e.u. of base concentration $= 2.5\%$. 1 e.u. of temperature $= 12.5\,°C$. 1 e.u. of humidity $= 12.5\%$ RH.

The signs of these interactions are determined by normal algebraic rules, and the overall design in coded data is shown in Table 6.8. Note that each column has an equal number of plus and minus signs. This is a useful check to ascertain whether signs have been correctly allocated.

It is often useful to depict three factor designs as a cube (Figure 6.5). All experiments with a high level of Factor X_1 [(+1, −1, −1), (+1, +1, −1), (+1, −1, +1), (+1, +1, +1)] appear on the right-hand face of the cube and all with a low level of the same factor [(−1, −1, −1), (−1, +1, −1), (−1, −1, +1), (−1, +1, +1)] on the left-hand face. Similarly, the high and low levels of Factor X_2 are represented by the top and bottom faces of the cube, respectively. High and low levels of Factor X_3 are represented by the back and front faces, respectively.

The magnitudes of the main effects of the factors and their interactions can now be calculated. The method is the same as before. Thus, for Factor X_1, the magnitude is the mean of all experiments with a high level of Factor X_1 minus the mean of all those with a low level of the same factor.

Taking this information from Table 6.8, the magnitude of the effect of Factor X_1 is

$$\frac{1}{4}[(5.3+6.6+3.6+7.0)-(1.6+3.4+2.6+3.0)]=+2.975$$

Similarly, the magnitude of Interaction $X_1X_2X_3$ is the mean of all experiments with a positive value for $X_1X_2X_3$ minus the mean of all experiments with a negative value for $X_1X_2X_3$. Therefore, from Table 6.8, the effect of Interaction $X_1X_2X_3$ is

$$\frac{1}{4}[(5.3+3.4+2.6+7.0)-(1.6+6.6+3.6+3.0)]=+0.875$$

TABLE 6.8
Signs to Calculate Main Effects and Interactions of a Three-Factor, Two-Level Factorial Design

Experiment	Factor			Interaction				Tablet Color
	X_1	X_2	X_3	X_1X_2	X_1X_3	X_2X_3	$X_1X_2X_3$	
(−1, −1, −1)	−1	−1	−1	+1	+1	+1	−1	1.6
(+1, −1, −1)	+1	−1	−1	−1	−1	+1	+1	5.3
(−1, +1, −1)	−1	+1	−1	−1	+1	−1	+1	3.4
(+1, +1, −1)	+1	+1	−1	+1	−1	−1	−1	6.6
(−1, −1, +1)	−1	−1	+1	+1	−1	−1	+1	2.6
(+1, −1, +1)	+1	−1	+1	−1	+1	−1	−1	3.6
(−1, +1, +1)	−1	+1	+1	−1	−1	+1	−1	3.0
(+1, +1, +1)	+1	+1	+1	+1	+1	+1	+1	7.0

Note: Center point (0, 0, 0): base concentration = 2.5%, temperature = 32.5 °C, humidity = 62.5% RH. 1 e.u. of base concentration = 2.5%. 1 e.u. of temperature = 12.5 °C. 1 e.u. of humidity = 12.5% RH.

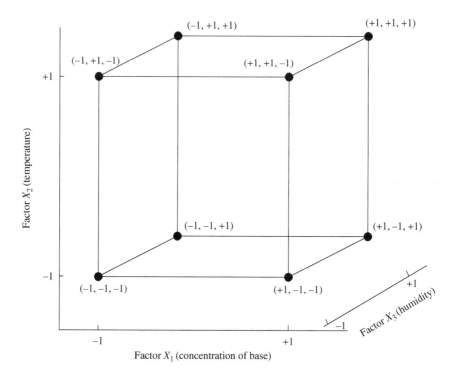

FIGURE 6.5   A three-factor, two-level experimental design.

TABLE 6.9
Magnitudes of the Main Effects and Interactions of the Factors Given in Table 6.8

Factor			Interaction			
X_1	X_2	X_3	X_1X_2	X_1X_3	X_2X_3	$X_1X_2X_3$
+2.975	+1.725	−0.175	+0.625	−0.475	+0.175	+0.875

The complete set of values for main effects and interactions is given in Table 6.9. A graphical representation such as Figure 6.6 is useful for visualizing the relative importance of these.

The most important factors are the concentration of base and the storage temperature. The environmental humidity is of less importance, as are all interactions. These conclusions, however, are based on a subjective assessment of the values shown in Table 6.9.

6.5   FACTORIAL DESIGN AND ANALYSIS OF VARIANCE

Factorial design becomes an even more powerful technique when allied to analysis of variance (ANOVA), because then an objective rather than a subjective

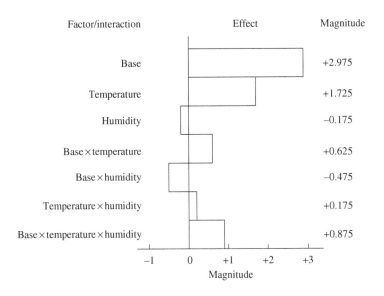

FIGURE 6.6 Graphical representation of the magnitudes of main effects and interactions of the factors shown in Table 6.9.

assessment of the relative importance of the various factors and interactions can be obtained.

6.5.1 YATES'S TREATMENT

A useful technique for linking factorial design to ANOVA was first described by Yates[3] and is best demonstrated by using the same worked example as before.

The experimental data are first tabulated in standard order. Then, the first two results relating to experiments (−1, −1, −1) and (+1, −1, −1) are added together (1.6 and 5.3) and the result (6.9) is put into in the first row of the column headed "Column 1" of Table 6.10a. The next two figures, relating to experiments (−1, +1, −1) and (+1, +1, −1), are then added together (3.4+6.6= 10) and the result is put into the second row of Column 1. The next two pairs (2.6+3.6=6.2, 3.0+7.0= 10) are treated similarly. Then, the difference between the first two experiments (+1, −1, −1) and (−1, −1, −1) is calculated (5.3−1.6=3.7) and the result placed in the fifth row of Column 1. The remaining three adjacent pairs are treated identically. At this stage, Table 6.10a appears as shown.

The process is then repeated using the numbers in Column 1, and the results are placed into Column 2. Thus, the first number in Column 2 is 16.9, obtained by adding together the first two rows in Column 1, namely, 6.9 and 10.0. The difference between these two numbers, 3.1, forms the fifth row of Column 2. The identical process is repeated yet again on the numbers in Column 2, and the results are placed in Column 3 (Table 6.10b).

Column 3 is now divided by 2^{N-1}, where N is the number of factors examined (in this case, 3). Therefore, $2^{N-1}=4$. These results, the average effects, are put into Column 4. Lastly, the mean squares are obtained by squaring the numbers in Column 3

TABLE 6.10A
Commencement of Yates's Treatment of Data from a Three-Factor, Two-Level Factorial Experimental Design

Experiment	Tablet Color	Column 1
(−1, −1, −1)	1.6	6.9
(+1, −1, −1)	5.3	10.0
(−1, +1, −1)	3.4	6.2
(+1, +1, −1)	6.6	10.0
(−1, −1, +1)	2.6	3.7
(+1, −1, +1)	3.6	3.2
(−1, +1, +1)	3.0	1.0
(+1, +1, +1)	7.0	4.0

Note: The tablet color data is taken from Table 6.7.

TABLE 6.10B
The Second Stage in Yates's Treatment of Data from a Three-Factor, Two-Level Factorial Experimental Design

Experiment	Tablet Color	Column 1	Column 2	Column 3
(−1, −1, −1)	1.6	6.9	16.9	—
(+1, −1, −1)	5.3	10.0	16.2	11.9
(−1, +1, −1)	3.4	6.2	6.9	6.9
(+1, +1, −1)	6.6	10.0	5.0	2.5
(−1, −1, +1)	2.6	3.7	3.1	−0.7
(+1, −1, +1)	3.6	3.2	3.8	−1.9
(−1, +1, +1)	3.0	1.0	−0.5	0.7
(+1, +1, +1)	7.0	4.0	3.0	3.5

and dividing by 2^N. Thus, the mean square attributable to experiment (+1, −1, −1) is $11.9^2/8 = 17.70$. The mean squares are put into Column 5, and the table now becomes Table 6.10c.

The importance of listing the experiments in standard order should now be apparent. Also, the values in Column 4 are those of the main effects and interactions first shown in Table 6.9.

The mean squares can now be placed in an ANOVA table (Table 6.11). In any factorial of the form 2^N, each effect and interaction has 1 degree of freedom. It remains to calculate F, the ratio between the mean squares and the residual squares, also known as the error mean square.

If the whole experiment had been replicated, then more than one observation would be available for each experiment and hence an estimate of the experimental error could be made. This is undoubtedly the favored approach and is dealt with

TABLE 6.10C
The Final Stage in Yates's Treatment of Data from a Three-Factor, Two-Level Factorial Experimental Design

Experiment	Tablet Color	Column 1	Column 2	Column 3	Column 4	Column 5
(−1, −1, −1)	1.6	6.9	16.9	—	—	—
(+1, −1, −1)	5.3	10.0	16.2	11.9	2.975	17.70
(−1, +1, −1)	3.4	6.2	6.9	6.9	1.725	5.95
(+1, +1, −1)	6.6	10.0	5.0	2.5	0.625	0.78
(−1, −1, +1)	2.6	3.7	3.1	−0.7	−0.175	0.06
(+1, −1, +1)	3.6	3.2	3.8	−1.9	−0.475	0.45
(−1, +1, +1)	3.0	1.0	−0.5	0.7	0.175	0.06
(+1, +1, +1)	7.0	4.0	3.0	3.5	0.875	1.53

TABLE 6.11
Analysis of Variance Table Following Yates's Treatment of Data Originally Shown in Table 6.7

Factor or Interaction	Experiment	Degrees of Freedom	Mean Square	F
Base	(+1, −1, −1)	1	17.70	295
Temperature	(−1, +1, −1)	1	5.95	99
Humidity	(−1, −1, +1)	1	0.06	—
Base × temperature	(+1, +1, −1)	1	0.78	13
Base × humidity	(+1, −1, +1)	1	0.45	7
Temperature × humidity	(−1, +1, +1)	1	0.06	—
Base × temperature × humidity	(+1, +1, +1)	1	1.53	25

later. However, replication may lead to an unacceptably high number of experimental runs. In these circumstances, the usual approach is to assume that some interactions have a negligible effect, and hence experimental runs containing these can be combined to give the experimental error. Alternatively, results that give very low values in the mean squares column may be combined for this purpose. Some caution is necessary here, in that incorrect assumptions can be made, and factors and interactions that are truly significant are assumed to be zero. Knowledge of the experimental system being studied and the use of common sense will help select those interactions that are likely to be of least significance. Adopting this approach for the information shown in Table 6.11, the mean squares relating to experiments (−1, −1, +1) and (−1, +1, +1) are distinctly lower than the others. These can therefore be combined to give a mean of 0.06, as the experimental error of the system, and F is calculated by dividing the other mean squares by this number. Table 6.11 gives the complete ANOVA table.

The significance of the values of F is assessed by comparing them with tabulated values. The numerator has 1 degree of freedom and the denominator has 2. Therefore, for $P < 0.05$, F should exceed 18.5. For $P < 0.01$, F should be greater than 98.5. Thus the presence of base is clearly the most important factor.

6.5.2 FACTORIAL DESIGN AND LINEAR REGRESSION

Yates's method of applying ANOVA to factorial design involves tedious albeit simple calculations. Evaluating mean effects and interactions resulting from a factorially designed experiment by application of multiple regression involves more complex calculations. However, the ready availability of computing power has removed both the tedium and the complexity of the calculations, and hence the use of regression is now overwhelmingly the method of choice. This is a particularly useful technique because it is the basis of relating factorial design to response-surface methodology and will be used extensively in later chapters.

It is essential to transform the numerical values of the factors by coding before regression is carried out. The coded factors are now represented by x_1, x_2, and x_3, and the response, tablet color, by y. Therefore, an equation which takes into account all the main effects and interactions is given by (6.4)

$$y = \beta_0 + \beta_1 x_1 + \beta_2 x_2 + \beta_3 x_3 + \beta_{12} x_1 x_2 + \beta_{13} x_1 x_3 + \beta_{23} x_2 x_3 + \beta_{123} x_1 x_2 x_3 \qquad (6.4)$$

where
β_0, β_1, and so on = the coefficients of the various terms in the equation.

The values of the coefficients of (6.4) are determined by multiple linear regression analysis. They are shown in Table 6.12 and in graphical form in Figure 6.7. Main effects represent the average result of changing one factor from -1 to $+1$, and the interactions show the result when any two or all three factors are changed simultaneously. If none of the factors had had any effect, then the responses, that is, tablet color, would be scattered randomly around their mean value, 4.1. Approximately the same value, 4.138, is obtained for b_0 when values of zero for x_1, x_2, and x_3 are substituted into (6.4).

TABLE 6.12
Regression Coefficients
Corresponding to all Main
Factors and Interactions

Mean	$b_0 = 4.138$
Main effects	$b_1 = 1.486$
	$b_2 = 0.863$
	$b_3 = -0.088$
Interactions	$b_{12} = 0.312$
	$b_{13} = -0.238$
	$b_{23} = 0.088$
	$b_{123} = 0.438$

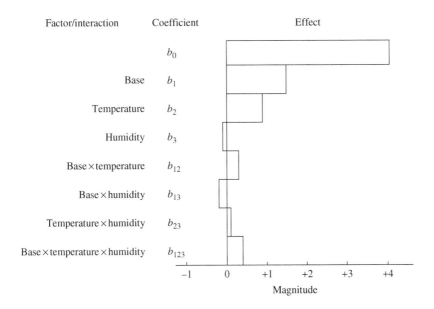

FIGURE 6.7 Graphical representation of the magnitudes of main effects and interactions of the factors shown in Table 6.12, calculated by linear regression.

From the results, it is apparent that only b_1, b_2, b_{12}, b_{13}, and b_{123} are significant. Hence, the dominant effects in the experiment are exerted by Factor X_1, Factor X_2, and Interactions X_1X_2, X_1X_3, and $X_1X_2X_3$. To support this conclusion, the coefficients of an equation (6.5) involving only these factors are calculated by multiple linear regression

$$y = \beta_0 + \beta_1 x_1 + \beta_2 x_2 + \beta_{12} x_1 x_2 + \beta_{13} x_1 x_3 + \beta_{123} x_1 x_2 x_3 \qquad (6.5)$$

These values are given in Table 6.13. The values of the coefficients quoted in Table 6.13 are virtually identical to the corresponding values in Table 6.12. The overall correlation coefficient of 0.9977 indicates goodness of fit.

TABLE 6.13
Regression Coefficients Corresponding to all Significant Main Factors and Interactions

Mean	$b_0 = 4.138$
Main effects	$b_1 = 1.488$
	$b_2 = 0.863$
Interactions	$b_{12} = 0.313$
	$b_{13} = -0.238$
	$b_{123} = 0.438$
Correlation coefficient	$r = 0.9977$

TABLE 6.14
Significant Main Factors and Interactions

Main Effects		Interactions		
X_1	X_2	X_1X_2	X_1X_3	$X_1X_2X_3$
2.97	1.72	0.62	−0.47	0.87

With the exception of the constant term b_0, the values of the regression coefficients are half the value of those shown in Table 6.9. This is because each coefficient in the latter table measures the change in response over a change in the corresponding factor of two units (-1 to $+1$). The significant main effects and interactions are shown in Table 6.14, and these are virtually identical to those obtained by the Yates treatment. The application of linear regression to factorial design has been fully discussed by Strange[4] and Gonzalez.[5]

6.6 REPLICATION IN FACTORIAL DESIGNS

As described in Chapter 4, calculation of the eight coefficients in (6.4) is carried out by solving eight simultaneous equations. Each of the latter is derived from one of the experiments by substituting the values of the factors of that experiment into (6.4). Thus, taking the first experiment in Table 6.8 as an example, (6.4) becomes (6.6)

$$1.6 = b_0 + (b_1 \times -1) + (b_2 \times -1) + (b_3 \times -1) + (b_{12} \times +1)$$
$$+ (b_{13} \times +1) + (b_{23} \times +1) + (b_{123} \times -1) \tag{6.6}$$

The other seven experiments in the design give similar equations, solution of which gives the coefficients.

The design shown in Table 6.7 is described as a saturated design, because there are as many experiments (8) as there are coefficients in the model equation (6.4). No experiment is repeated. Hence, there is no way of assessing whether a measured change in the response is brought about by changing the values of the factors or whether it is due to experimental error in measuring the response.

The regression equation (6.4) is a simplification, because in any experimental procedure there will always be uncontrolled factors or errors associated with measurement, so the measured response will vary randomly even if all the factors are kept constant. To account for this, an extra term (ε) is introduced into (6.4) to give (6.7)

$$y = \beta_0 + \beta_1 x_1 + \beta_2 x_2 + \beta_3 x_3 + \beta_{12} x_1 x_2 + \beta_{13} x_1 x_3 + \beta_{23} x_2 x_3 + \beta_{123} x_1 x_2 x_3 + \varepsilon \tag{6.7}$$

A saturated design cannot give an estimate of ε, and hence, the significance of the coefficients cannot be measured by statistical techniques. Nevertheless, a fundamental property of all factorial designs is that all the coefficients are estimated with equal precision. Furthermore, all coefficients have the same units, namely, that of the response, because they are coefficients of the coded variables which are dimensionless. Therefore, provided coding has taken place, they can be directly compared with each other.

To assess the underlying experimental error, some degree of replication must occur. One way of achieving this would be to repeat the whole experimental design, thus doubling the number of experiments. This will be described using the example given earlier involving the discoloration of lactose tablets. Every experiment is performed twice, and Table 6.15a shows the experimental design as before and two sets of responses.

The means of the eight pairs of duplicates are calculated, then the difference of each result from that mean, and hence the variance. The total variance is 0.160 and is for 16 experiments. There are $16-1=15$ degrees of freedom. Each factor and interaction has 1 degree of freedom, and hence, there are $15-7=8$ degrees of freedom for the experimental error.

It is convenient for the application of Yates's treatment to carry out the calculations based on the total of each pair of duplicated results, which will have 1 degree of freedom each. The various stages of the Yates treatment are summarized in Table 6.15b, which is similar in structure to Table 6.10c. The corresponding ANOVA table (Table 6.16) is similar to Table 6.11.

The mean square of each factor and interaction is divided by the mean square of the experimental error to give F. In this way, a value for F is obtained for each factor and interaction, rather than assuming that some of these were negligible and therefore could be used as a substitute for experimental error.

Regression can also be applied to the 16 values of tablet color given in Table 6.15a, giving (6.8)

$$Y=4.175+1.538X_1+0.925X_2-0.125X_3+0.288X_1X_2$$
$$-0.238X_1X_3+0.1X_2X_3+0.438X_1X_2X_3 \qquad (6.8)$$

TABLE 6.15A
A Duplicated Three-Factor, Two-Level Experimental Design

| Experiment | Tablet Color | | Mean | $(Xm - X)$ | $\Sigma(Xm - X)_2/(n - 1)$ |
	Set 1	Set 2			
$(-1, -1, -1)$	1.6	1.5	1.55	0.05	0.005
$(+1, -1, -1)$	5.3	5.5	5.40	0.10	0.020
$(-1, +1, -1)$	3.4	3.6	3.50	0.10	0.020
$(+1, +1, -1)$	6.6	6.9	6.75	0.15	0.045
$(-1, -1, +1)$	2.6	2.3	2.45	0.15	0.045
$(+1, -1, +1)$	3.6	3.6	3.60	0.00	0.000
$(-1, +1, +1)$	3.0	3.1	3.05	0.05	0.005
$(+1, +1, +1)$	7.0	7.2	7.10	0.10	0.020
Total					0.160

TABLE 6.15B
The Final Stage in Yates's Treatment of Data from a Duplicated Three-Factor, Two-Level Experimental Design

Experiment	Tablet Color							
	Set 1	Set 2	Total	Column 1	Column 2	Column 3	Column 4	Column 5
$(-1, -1, -1)$	1.6	1.5	3.1	13.9	34.4	—	—	—
$(+1, -1, -1)$	5.3	5.5	10.8	20.5	32.4	24.6	6.15	37.87
$(-1, +1, -1)$	3.4	3.6	7.0	12.1	14.2	14.8	3.70	13.69
$(+1, +1, -1)$	6.6	6.9	13.5	20.3	10.4	4.6	1.15	1.32
$(-1, -1, +1)$	2.6	2.3	4.9	7.7	6.6	-2.0	-0.5	0.75
$(+1, -1, +1)$	3.6	3.6	7.2	6.5	8.2	-3.8	-0.95	0.90
$(-1, +1, +1)$	3.0	3.1	6.1	2.3	-1.2	1.6	0.40	0.16
$(+1, +1, +1)$	7.0	7.2	14.2	8.1	5.8	7.0	1.75	3.06

TABLE 6.16
Analysis of Variance Table Following Yates's Treatment of the Data Originally Shown in Table 6.15a

Factor or Interaction	Experiment	Degrees of Freedom	Mean Square	F
Base	$(+1, -1, -1)$	1	37.87	1893
Temperature	$(-1, +1, -1)$	1	13.69	684
Humidity	$(+1, +1, -1)$	1	0.25	12
Base × temperature	$(-1, -1, +1)$	1	1.32	66
Base × humidity	$(+1, -1, +1)$	1	0.90	45
Temperature × humidity	$(-1, +1, +1)$	1	0.16	8
Base × temperature × humidity	$(+1, +1, +1)$	1	3.06	153
Experimental error		8	0.02	

The correlation coefficient of this equation is 0.9986.

This approach was adopted by Plazier-Vercammen and De Neve[6] in a study of complex formation by povidone with salicylic and benzoic acids.

In this context, replication means carrying out the whole experimental design on more than one occasion. It must not be confused with replicated measurements of the response from the same experiment.

Complete replication of the whole design may not be feasible because of the extra work that it would generate. A more usual method of assessing the underlying experimental error in the measurement of the response is to replicate only some of the experiments. The center point of the design is often chosen for this purpose. Thus, for a three-factor, two-level design, the center point is designated $(0, 0, 0)$. For the data described in Table 6.7, the values of the three factors at the center point are $X_1 = 2.5\%$, $X_2 = 32.5\,°C$, and $X_3 = 62.5\%$ RH. This has the effect of introducing a third level and will be discussed more fully when three-level designs are introduced later in this chapter.

6.7 THE SEQUENCE OF EXPERIMENTS

As mentioned earlier, all experiments in a design should ideally be carried out simultaneously, using the same equipment and personnel. In this way uncontrolled factors, both known and unknown, will affect each experiment equally. This is rarely possible. The most common solution to this problem is to carry out the experiments in a random order. Hence, uncontrolled variables will affect the responses in a random manner. However, before a randomized design is embarked upon, there must be confidence that the response to all experiments in the design can be measured with the same precision.

For example, the crushing strength of tablets is increased by raising the compression pressure and is reduced in the presence of a lubricant. Thus, in an experimental design to study the effects of pressure and lubricant on tablet strength, the experiment in which pressure is highest and lubricant content lowest would be expected to produce the strongest tablets. All equipment used to measure crushing strength has an upper limit; therefore, it would be prudent to carry out the experiment that gives the strongest tablets first. If the testing apparatus can accommodate these tablets, then one can be confident that tablets from all the other experiments in the design can be measured. If they cannot, the design can be modified at an early stage.

An alternative to randomization is possible if the response is believed to change uniformly with time. If a three-factor, two-level experiment is carried out in standard order, and each experiment is separated by the same time interval (t), then the relationship between the measured response (y') and the true response (y) of the ith experiment is given by (6.9)

$$y'_i = y_i + (i-1)t \tag{6.9}$$

Thus, the observed response for the second experiment will be $y_2 + t$ and for the eighth experiment $y_8 + 7t$.

If the experiments are carried out in standard order, responses altered by a time trend will affect the calculated magnitudes of all main effects. Thus, referring to Figure 6.5, the effect of Factor X_1 will be

$$\frac{1}{4}\left\{\left[(y_2+t)+(y_4+3t)+(y_6+5t)+(y_8+7t)\right]-\left[(y_1)+(y_3+2t)+(y_5+4t)+(y_7+6t)\right]\right\}$$

$$=\frac{1}{4}\left[(y_2+y_4+y_6+y_8+16t)-(y_1+y_3+y_5+y_7+12t)\right]$$

As this equation is further simplified, the values of t do not cancel out, and therefore the response to this factor will apparently be increased by $1/4 \times 4t$. However, in the calculation of the three-way interaction $X_1X_2X_3$, all t's are eliminated, and hence the

TABLE 6.17
Three-Factor, Two-Level Factorial Design with a Time Trend

Experiments in Standard Order	Observed Response (y')	Rearranged Order	Observed Response (y')
(−1, −1, −1)	y_1	(−1, +1, +1)	y_7
(+1, −1, −1)	$y_2 + t$	(+1, +1, −1)	$y_4 + t$
(−1, +1, −1)	$y_3 + 2t$	(+1, −1, +1)	$y_6 + 2t$
(+1, +1, −1)	$y_4 + 3t$	(−1, −1, −1)	$y_1 + 3t$
(−1, −1, +1)	$y_4 + 4t$	(+1, −1, −1)	$y_2 + 4t$
(+1, −1, +1)	$y_6 + 5t$	(−1, −1, +1)	$y_5 + 5t$
(−1, +1, +1)	$y_7 + 6t$	(−1, +1, −1)	$y_3 + 6t$
(+1, +1, +1)	$y_8 + 7t$	(+1, +1, +1)	$y_8 + 7t$

time trend has no effect. It is always desirable for the main effects to be independent of the time trend rather than the interactions, and it is possible to rearrange the order of the experiments to achieve this. The revised order is shown in Table 6.17.

Factor X_1 now becomes

$$\frac{1}{4}\left\{\left[(y_2 + 4t) + (y_4 + t) + (y_6 + 2t) + (y_8 + 7t)\right] - \left[(y_1 + 3t) + (y_3 + 6t) + (y_5 + 5t) + (y_7)\right]\right\}$$

$$= \frac{1}{4}\left[(y_2 + y_4 + y_6 + y_8 + 14t) - (y_1 + y_3 + y_5 + y_7 + 14t)\right]$$

and all the t's cancel out. The complete list of responses is given in Table 6.17.

6.8 FACTORIAL DESIGNS WITH THREE LEVELS

The applications of factorial analysis so far described deal with only two levels of a particular factor, for example, low and high, or −1 or +1. If there are only two points, they can only be joined by a straight line. This implies that there is a rectilinear relationship between the magnitude of the factor and the response. If this assumption is not true, then a maximum or minimum value of the response may occur between the chosen levels of the factors and this would not be detected. Therefore, if a rectilinear relationship cannot be safely assumed, then it is necessary to use more than two levels.

The numerical notation used earlier in this chapter can be extended to three or more levels. If a third level is to be used, it is usually, though not always, set equidistant from the lower (−1) to higher (+1) levels and is designated 0. Obviously, the alternative alphabetical notation previously described can only be used for two-level studies and hence is inappropriate here.

Factors are usually designated by capital letters X_1, X_2, and so on, as before, and also as before, it is often useful to envisage the experimental design in diagrammatic form. Thus, Figure 6.8 represents a two-factor, three-level experimental design.

The use of three levels implies the possibility of nonlinear relationships between the two factors and the response. The full regression equation (6.10) contains linear terms (X_n to the power 1) and quadratic terms (X_n to the power 2) and all interactions between the factors.

$$y = \beta_0 + \beta_1 X_1 + \beta_{11} X_1^2 + \beta_2 X_2 + \beta_{22} X_2^2 + \beta_{12} X_1 X_2 + \varepsilon \qquad (6.10)$$

As before, this will be demonstrated by a worked example. A metered-dose inhaler delivers droplets to the lung with a wide spectrum of sizes. Only some of these droplets can be deposited in the required part of the respiratory tract, and these form the respirable fraction, expressed as a percentage. The magnitude of this fraction is governed by many factors. Two of these are the concentration of surfactant in the system and the concentration of water. A third important factor is the aperture of the valve on the pack, and this is introduced into the discussion later. However, for the present example, it will be assumed that the same design of valve is used throughout.

The objective of the experiment is therefore to determine the effect of surfactant and water concentrations on the respirable fraction delivered by the device. Surfactant concentration is designated Factor X_1, and three levels (0.5, 1.0, and 1.5%) are chosen. These are designated levels −1, 0, and +1, respectively. Water concentration is designated Factor X_2, and the levels are 1.4, 2.8, and 4.2%, and these too are

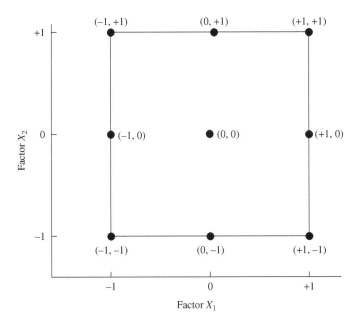

FIGURE 6.8 A two-factor, three-level experimental design.

designated -1, 0, and $+1$, respectively. The design of the experiment is shown in Table 6.18, and the values of the factors are given in experimental units.

The nine experiments are carried out in random order, and the results are shown in Table 6.18. Data are analyzed as described under *Two-Way Analysis of Variance* in Chapter 2.

1. The total of every column, row, and the grand total (397.5) is calculated.
2. The "correction term" $= 397.5^2/9 = 17{,}556.25$.
3. The total sum of squares minus the correction term

$$= (52.5^2 + 50.0^2 + \cdots + 24.1^2) - 17{,}556.25 = 18{,}461.51 - 17{,}556.25 = 905.26$$

4. The sum of squares of Factor X_1 minus the correction term

$$= \frac{162.0^2 + 139.2^2 + 96.3^2}{3} - 17{,}556.25 = 18{,}298.11 - 17{,}556.25 = 741.86$$

5. The sum of squares of Factor X_2 minus the correction term

$$= \frac{141.6^2 + 134.2^2 + 121.7^2}{3} - 17{,}556.25 = 17{,}623.70 - 17{,}556.25 = 67.44$$

6. The residual sum of squares

$$= 905.26 - (741.86 + 67.44) = 95.96$$

The ANOVA table (Table 6.19) can now be constructed.

TABLE 6.18
Two-Factor, Three-Level Factorial Design to Investigate the Influence of Surfactant Concentration (Factor X_1) and Water Concentration (Factor X_2) on the Respirable Fraction (%) Obtained from a Metered-Dose Inhaler

	Factor X_1						Total
Factor X_2	-1		0		$+1$		
-1	$(-1, -1)$	52.5	$(0, -1)$	50.0	$(+1, -1)$	39.1	141.6
0	$(-1, 0)$	53.2	$(0, 0)$	47.9	$(+1, 0)$	33.1	134.2
$+1$	$(-1, +1)$	56.3	$(0, +1)$	41.3	$(+1, +1)$	24.1	121.7
Total		162.0		139.2		96.3	397.5

Note: Center point $(0, 0)$: surfactant concentration $= 1\%$, water concentration $= 2.8\%$. 1 e.u. of surfactant concentration $= 0.5\%$. 1 e.u. of water concentration $= 1.4\%$.

TABLE 6.19
Analysis of Variance Table of Data Given in Table 6.18

Source	Sum of Squares	Degrees of Freedom	Mean Square
X_1	741.86	2	370.93
X_2	67.45	2	33.73
X_1X_2	95.95	4	23.99
Total	905.26	8	—

Both Factors X_1 and X_2 have linear and quadratic terms, and hence analysis can now be taken further. The responses are multiplied by the coefficients given in Table 6.20. The derivation of these coefficients is beyond the scope of this book, but appropriate references for further study are given in the Further Reading section. Consider the first row of coefficients in Table 6.20. This refers to the linear effect of Factor X_1 and compares the responses at the three lowest levels of Factor X_1, that is, experiments (−1, −1), (−1, 0), (−1, +1), with responses at the three highest levels of Factor X_1 [(+1, −1), (+1, 0), (+1, +1)], taken across all levels of Factor X_2. The coefficients of the interaction terms are obtained by multiplying together those of the main effects. These coefficients are now applied to the responses given in Table 6.18.

Thus,

$$\begin{aligned} X_1 = &(52.5\times-1)+(53.2\times-1)+(56.3\times-1)+(50.0\times0)+(47.9\times0)\\ &+(41.3\times0)+(39.1\times+1)+(33.1\times+1)+(24.1\times+1)=-65.7 \end{aligned}$$

The corresponding sum of squares is $-65.7^2/6=719.4$.

By identical methods, the multiples of the other responses and coefficients and sums of squares are calculated. They are summarized in Table 6.21, which also contains ANOVA data.

TABLE 6.20
Coefficients for a Two-Factor, Three-Level Factorial Design

	Treatment Combination Coefficients									Sum of Squared
Factor	**(−1, −1)**	**(−1, 0)**	**(−1, +1)**	**(0, −1)**	**(0, 0)**	**(0, +1)**	**(+1, −1)**	**(+1, 0)**	**(+1, +1)**	**Coefficients**
X_1	−1	−1	−1	0	0	0	+1	+1	+1	6
X_1^2	+1	+1	+1	−2	−2	−2	+1	+1	+1	18
X_2	−1	0	+1	−1	0	+1	−1	0	+1	6
X_2^2	+1	−2	+1	+1	−2	+1	+1	−2	+1	18
X_1X_2	+1	0	−1	0	0	0	−1	0	+1	4
$X_1X_2^2$	−1	+2	−1	0	0	0	+1	−2	+1	12
$X_1^2X_2$	−1	0	+1	+2	0	−2	−1	0	+1	12
$X_1^2X_2^2$	+1	−2	+1	−2	+4	−2	+1	−2	+1	36

TABLE 6.21
Sum of Squares and ANOVA for Data from Table 6.18

Source		Response × Coefficient	Sum of Squares	Degrees of Freedom	Mean Squares	F
X_1			741.8	2	370.9	
	X_1	−65.7	719.4	1	719.4	319
	X_1^2	−20.1	22.4	1	22.4	10
X_2			67.4	2	33.7	
	X_2	−19.9	66.0	1	66.0	29
	X_2^2	−5.1	1.4	1	1.4	
X_1X_2			96.0	4	24.0	
	X_1X_2	−8.8	88.4	1	88.4	39
	$X_1X_2^2$	−5.4	2.4	1	2.4	
	$X_1^2X_2$	6.2	3.2	1	3.2	
	$X_1^2X_2^2$	8.4	2.0	1	2.0	

The absence of replication precludes a proper calculation of the underlying error of the system. However, if the smallest mean squares (X_2^2, $X_1^2X_2$, $X_1X_2^2$, $X_1^2X_2^2$) are averaged, this can form the denominator of the F ratio. These F values are given in Table 6.21. Thus, X_1, X_1^2, X_2, and X_1X_2 are all significant at $P=0.05$, and of these, all except X_1^2 are significant at $P=0.01$. From this, it can be inferred that because all quadratic terms of the main factors and interactions are of low significance, a reasonably rectilinear relationship links both factors and the response, though interaction between the main effects is significant.

The Yates treatment can be applied to three-level factorial designs. Table 6.22 shows the standard order for a two-factor, three-level design, and the data in the response column are the same as those shown in Table 6.18.

The entries in the column headed "Column 1" are derived as follows. The first number in Column 1 is the sum of the first three responses, that is, experiments (−1, −1), (0, −1), and (+1, −1). Items 2 and 3 of this column are the sums of experiments (−1, 0), (0, 0), and (+1, 0) and (−1, +1), (0, +1), and (+1, +1), respectively.

TABLE 6.22
Yates's Treatment Applied to a Two-Factor, Three-Level Design, Using Data from Table 6.18

Experiment	Response	Column 1	Column 2	Effect	Divisor	Mean Square
(−1, −1)	52.5	141.6	397.5	—	—	—
(0, −1)	50.0	134.2	−65.7	X_1	6	719.4
(+1, −1)	39.1	121.7	−20.1	X_1^2	18	22.4
(−1, 0)	53.2	−13.4	−19.9	X_2	6	66.0
(0, 0)	47.9	−20.1	−18.8	X_1X_2	4	88.4
(+1, 0)	33.1	−32.2	6.2	$X_1^2X_2$	12	3.2
(−1, +1)	56.3	−8.4	−5.1	X_2^2	18	1.4
(0, +1)	41.3	−9.5	−5.4	$X_1X_2^2$	12	2.4
(+1, +1)	24.1	−2.2	8.4	$X_1^2X_2^2$	36	2.0

The fourth number in Column 1 is the difference between the first row of Column 1 and the third row of Column 1, that is, the response to experiment $(+1, -1)$ minus that of experiment $(-1, -1)$. The fifth number in this column is the difference between experiments $(+1, 0)$ and $(-1, 0)$, and the sixth the difference between $(+1, +1)$ and $(-1, +1)$. This process computes the linear component of the effect. The last third of the column is obtained by calculating the sum of the first and third items in each group of three minus twice the middle item of the group. This computes the quadratic component of that effect. Thus, the last number in Column 1 is given by $56.3+24.1-(2\times41.3)=-2.2$. The numbers in Column 2 are derived from those in Column 1 in the same way. The effects to which they relate are shown in the column headed "Effect."

The entries in the "Divisor" column are derived from the formula

$$\text{divisor}=2^r3^t n$$

where

$r=$ the number of factors in the effect

$t=$ the number of factors in the experiment minus the number of linear terms in this effect

$n=$ the number of replicates (in this case, 1).

The sum of squares is obtained by squaring each item in Column 2 and dividing by the corresponding entry in the divisor column. For example, the entry in the divisor column for the last term is 36. There are two factors in this effect $(X_{12}X_{22})$, there are two factors in this experiment but no linear terms, and the number of replicates is 1. The divisor is therefore $2^2\times3^2\times1=36$. Consequently, the last term in the mean square column is $8.4^2/36=1.96$, which is rounded to 2.0.

An ANOVA table can now be constructed (Table 6.23) and the results analyzed as before by calculating F.

TABLE 6.23
Analysis of Variance Table for the Data in Table 6.22

Source of Variation		Sum of Squares	Degrees of Freedom	Mean Square	F
X_1		741.8	2	370.9	
	X_1	719.4	1	719.4	319
	X_1^2	22.4	1	22.4	10
X_2		67.4	2	33.7	
	X_2	66.0	1	66.0	29
	X_2^2	1.4	1	1.4	
X_1X_2		96.0.	4	24.0	
	X_1X_2	88.4	1	88.4	39
	$X_1X_2^2$	2.4	1	2.4	
	$X_1^2X_2$	3.2	1	3.2	
	$X_1^2X_2^2$	2.0	1	2.0	
Error			27		
Total		905.3	35		

Regression of the data in Table 6.18 gives (6.11), with a correlation coefficient of 0.9958.

$$y = 46.966 - 10.95X_1 - 3.350X_1{}^2 - 3.317X_2 - 0.850X_2{}^2 - 4.70X_1X_2 \qquad (6.11)$$

6.9 THREE-FACTOR, THREE-LEVEL FACTORIAL DESIGNS

The next stage in complexity is the three-factor design where each factor is studied at three levels. The experimental layout and notation are shown in Figure 6.9. As there are 27 possible combinations, there are 26 degrees of freedom. Each main effect has 2 degrees of freedom, the two-factor interactions have 4 degrees of freedom each, and the three-factor interaction has 8 degrees of freedom. If the factors are quantitative and equally spaced, the main effects can be partitioned into linear and quadratic components as before, as can the interaction terms. The eight possible combinations which can be derived from the three-way interaction ($X_1X_2X_3$, $X_1{}^2X_2X_3$, etc.) are often difficult to explain on a practical basis, and hence the $X_1X_2X_3$ interaction often serves as the "error" by which the main effects and two-way interactions are tested.

The example involving a metered-dose inhaler used earlier can usefully be extended into a three-factor, three-level design. The third factor to be introduced is

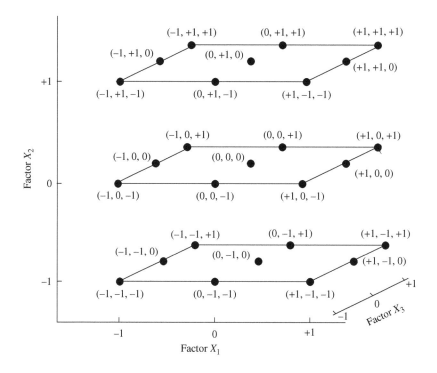

FIGURE 6.9 A three-factor, three-level experimental design.

the aperture of the valve. Thus, surfactant concentration is Factor X_1, water concentration Factor X_2, and valve aperture Factor X_3. The experimental design and data are shown in Table 6.24. The relative importance of the factors and interactions is calculated by ANOVA as before.

1. The total of every column, row, and the grand total (1034.4) is calculated.
2. The "correction term" $= (1034.4)^2/27 = 39,629.0$.
3. The next stage is to calculate the sums of squares of the main effects of Factors X_1, X_2, and X_3.
   Thus, for Factor X_1
   The sum of squares of all results when Factor X_1 is at level -1

$$= \frac{(52.5+53.2+56.3+46.2+53.8+33.5+40.3+38.6+29.4)^2}{9} = \frac{(403.8)^2}{9} = 18,117.2$$

Similarly, when Factor X_1 is at level 0, the sum of squares

$$= \frac{(50.0+47.9+\cdots+21.9)^2}{9} = \frac{(350.4)^2}{9} = 13,642.2$$

Similarly, when Factor X_1 is at level $+1$, the sum of squares

$$= \frac{(39.1+33.1+\cdots+15.0)^2}{9} = \frac{(280.2)^2}{9} = 8723.6$$

TABLE 6.24
Three-Factor, Three-Level Factorial Design to Investigate the Influence of Surfactant Concentration (Factor X_1), Water Concentration (Factor X_2), and Valve Aperture (X_3) on the Respirable Fraction (%) Obtained from a Metered Dose Inhaler

	Factor X_1 (Surfactant Concentration)									
	−1			0			+1			
	Factor X_2 (Water Concentration)									
Factor X_3 (Valve Aperture)	−1	0	+1	−1	0	+1	−1	0	+1	Total
−1	52.5	53.2	56.3	50.0	47.9	41.3	39.1	33.1	24.1	397.5
0	46.2	53.8	33.5	52.9	43.4	19.1	47.0	32.7	18.5	347.1
+1	40.3	38.6	29.4	39.7	34.2	21.9	40.2	30.5	15.0	289.8
Total	139.0	145.6	119.2	142.6	125.5	82.3	126.3	96.3	57.6	1034.4

Adding these three terms together and subtracting the correction term gives

$$(18,117.2+13,642.2+8723.6)-39,629.0=40,483.0-39,629.0=854.0$$

By identical methods, the main effects of Factors X_2 and X_3 can be calculated. For Factor X_2
The sum of squares when Factor X_2 is at level -1

$$=\frac{(407.9)^2}{9}=18,486.9$$

The sum of squares when Factor X_2 is at level 0

$$=\frac{(367.4)^2}{9}=14,998.1$$

The sum of squares when Factor X_2 is at level $+1$

$$=\frac{(259.1)^2}{9}=7459.2$$

Totaling these three terms and subtracting the correction term gives

$$(18,486.9+14,998.1+7459.2)-39,629.0=1315.2$$

For Factor X_3
The sum of squares when Factor X_3 is at level -1

$$=\frac{(397.5)^2}{9}=17,556.3$$

The sum of squares when Factor X_3 is at level 0

$$=\frac{(347.1)^2}{9}=13,386.5$$

The sum of squares when Factor X_3 is at level $+1$

$$=\frac{(289.8)^2}{9}=9331.6$$

Totaling these three terms and subtracting the correction term gives

$$(17{,}556.3 + 13{,}386.5 + 9331.6) - 39{,}629.0 = 40{,}274.4 - 39{,}629.0 = 645.4$$

4. The next stage is to calculate the three two-factor interactions X_1X_2, X_1X_3, and X_2X_3. For the X_1X_2 interaction, changes in Factor X_3 are ignored. For example, the results of experiments $(-1, -1, -1)$, $(-1, -1, 0)$, and $(-1, -1, +1)$ are added together and the sum squared. Because there are three terms, the sum of squares is divided by three, and the correction term subtracted. From this is then subtracted the sums of squares of the main effects of Factor X_1 and Factor X_2. What remains is the sum of the squares of the interaction X_1X_2.

   Thus, the sum of squares of the X_1X_2 interaction

$$= \frac{(52.5 + 46.2 + 40.3)^2}{3} + \frac{(53.2 + 53.8 + 38.6)^2}{3} + \cdots + \frac{(24.1 + 18.5 + 15.0)^2}{3}$$

$-39{,}629.0\,(\text{the correction term}) - (854.0 + 1315.2)\,(\text{the main effects of}$

Factors X_1 and X_2

$= 245.3$

   Similarly, to calculate the X_1X_3 interaction, changes in Factor X_2 are ignored. Thus, results from experiments $(-1, -1, -1)$, $(-1, 0, -1)$, and $(-1, +1, -1)$ are grouped together. The sum of squares for the X_1X_3 interaction

$$= \frac{(52.5 + 53.2 + 56.3)^2}{3} + \frac{(46.2 + 53.8 + 33.5)^2}{3} + \cdots + \frac{(40.2 + 30.5 + 15.0)^2}{3}$$

$-39{,}629.0\,(\text{the correction term}) - (854.0 + 645.4)\,(\text{the main effects of}$

Factors X_1 and X_3

$= 180.9$

   The sum of squares for the X_2X_3 interaction

$$= \frac{(52.5 + 50.0 + 39.1)^2}{3} + \frac{(46.2 + 52.9 + 47.0)^2}{3} + \cdots + \frac{(29.4 + 21.9 + 15.0)^2}{3}$$

$-39{,}629.0\,(\text{the correction term}) - (1315.2 + 645.4)\,(\text{the main effects of}$

Factors X_2 and X_3)

$= 297.1$

5. The next stage is to calculate the sum of squares of the three-way interaction $X_1X_2X_3$. This is done by calculating the sum of the squares of all the terms and subtracting from it the three main effects and the

TABLE 6.25
ANOVA Table for Data Presented in Table 6.24

Source	Sum of Squares	Degrees of Freedom	Mean Square	F
X_1	854.0	2	427.0	38.8
X_2	1315.2	2	657.6	59.8
X_3	645.4	2	322.7	29.3
X_1X_2	245.3	4	61.3	5.6
X_1X_3	180.9	4	45.2	4.1
X_2X_3	297.1	4	74.3	6.8
$X_1X_2X_3$	88.1	8	11.0	—
Total	3626.1	26	—	—

three two-way interactions. Thus, the sum of squares of the $X_1X_2X_3$ interaction

$$= (52.5^2+46.2^2+\cdots+15.0^2)-39{,}629.0 \text{ (the correction term)}$$
$$- (854.0+1315.2+645.4) \text{ (the three main effects)}$$
$$- (245.3+180.9+297.1) \text{ (the three two-way interactions)}$$
$$= 88.1$$

The ANOVA table can now be constructed (Table 6.25). In the absence of replication of individual data points, the $X_1X_2X_3$ interaction with its 8 degrees of freedom is used as the error term. Dividing all other mean squares by the mean square of the $X_1X_2X_3$ interaction gives the values of F shown in the right-hand column of Table 6.25. All main effects are significant at the 1% level of significance, but none of the interactions has significance even at the 5% level.

The Yates treatment can also be used in three-factor, three-level designs. In this case, the standard order is $(-1, -1, -1)$, $(0, -1, -1)$, $(+1, -1, -1)$, $(-1, 0, -1)$, $(0, 0, -1)$, $(+1, 0, -1)$, $(-1, +1, -1)$, $(0, +1, -1)$, $(+1, +1, -1)$, $(-1, -1, 0)$, $(0, -1, 0)$,..., $(+1, +1, +1)$.

Regression of the data given in Table 6.24 yields (6.12), with a correlation coefficient of 0.9511

$$y=41.7-6.866X_1-0.933X_1^2-8.269X_2-3.766X_2^2-6.433X_3-0.383X_3^2$$
$$-4.075X_1X_2+4.139X_1X_3-2.833X_2X_3+0.563X_1X_2X_3 \qquad (6.12)$$

6.9.1 MIXED OR ASYMMETRIC DESIGNS

It is not essential that all factors should be explored with the same number of levels. If it can be confidently assumed that a rectilinear relationship exists between the two extreme levels of the factor (i.e., −1 to +1) and a response, then a two-level study can be carried out using that factor, with more levels for the other factors. A good example of that approach is the work of Sanderson et al.[7] These workers assessed the relative importance of various formulation and process factors in the properties of paracetamol tablets. The factors examined at two levels were mixing time (1 min and 5 min), starch concentration (1% and 7%), and drug particle size (<20 μm and >20 μm). Three compression pressures were used (100, 150, and 200 MN·m^{-2}). A four-factor, three-level design would require 81 experiments (3^4), whereas the mixed design used by Sanderson needed 24 experiments plus any necessary replicates.

6.10 BLOCKED FACTORIAL DESIGNS

From the worked examples given earlier in this chapter, it is seen that as the number of factors and levels is increased, the number of experiments rises steeply, even if the experiments are not replicated. Thus, a two-factor, two-level design requires 4 experiments, a three-factor, two-level design requires 8 experiments, and a three-factor, three-level design requires 27 experiments. In general terms, if there are F factors and L levels, then L^F experiments are needed for a complete factorial design. Hence, the number of experiments can grow rapidly, and the consequent high consumption of time and materials may nullify the advantages of the factorial approach.

A further consequence of the large number of experiments is the difficulty in arranging for all experiments to be carried out simultaneously. One approach to minimize the impact of sequential rather than simultaneous experimentation is to group the experiments into blocks.

Consider the two-factor, two-level experiment described at the beginning of this chapter, in which the effects of temperature and catalyst concentration on the loss of Compound E were studied (Figure 6.4 and Table 6.5). For a complete design, four experiments are necessary, and these should ideally all be carried out at the same time. If only one set of apparatus is available, then the experiments must be carried out singly but in random order. However, consider the situation in which two sets of apparatus are available, so that the four experiments could be carried out in two pairs. It is possible to arrange the two pairs of samples in three ways. These are shown in Table 6.26.

Taking the first of these arrangements, experiments (−1, −1) and (+1, −1) are performed, followed by (−1, +1) and (+1, +1). In the first pair the catalyst concentration is low, and in the second pair the catalyst concentration is higher. Therefore, if catalyst concentration plays a role, the two pairs will be expected to differ. However, if the point in time at which the experiments were carried out also affects the results, then this latter effect cannot be separated from the effect of the catalyst. The two effects are said to be confounded.

TABLE 6.26
Possible Arrangement of Four Experiments into Two Pairs

	First Pair	Second Pair
Arrangement 1	(−1, −1), (+1, −1)	(−1, +1), (+1, +1)
Arrangement 2	(−1, −1), (−1, +1)	(+1, −1), (+1, +1)
Arrangement 3	(−1, −1), (+1, +1)	(+1, −1), (−1, +1)

By a similar argument, if the second arrangement is used, the effect of temperature would be confounded with that of time. The third arrangement is $(-1, -1)$ and $(+1, +1)$, followed by $(+1, -1)$ and $(-1, +1)$. In this case, neither main effect is confounded, but the interaction is. As a rule, main effects should not be confounded, and if confounding is unavoidable, it is better to confound an interaction, which of course may not play a significant role in any case.

This can be illustrated by further consideration of the experimental design described in Table 6.5. Using the data given there, and assuming that all experiments were carried out at the same time, the effect of Factor X_1 was found to be 15, that of Factor X_2 was 20, and the effect of the Interaction X_1X_2 was zero. Let us now suppose that the four experiments were then carried out in two pairs separated by a time interval, using the arrangements shown in Table 6.26. Also suppose that, unbeknown to the experimenter, some additional factor was operational when the second pair of experiments was performed; the effect of this additional factor is to increase the loss of E by an additional 10%.

The results of these experiments are shown in Table 6.27, together with the calculated values of the main effects and the interaction. It is seen that when a main effect is confounded, its value changes. Where the interaction is confounded, the values of the main effects remain unchanged, though that of the interaction changes.

In the above example, confounding is unavoidable because of constraints imposed by lack of equipment. However, confounding can often be used to advantage to reduce the number of experiments. If it can be decided at the outset that some interactions either do not occur or can safely be ignored, it is possible to run fewer combinations of factors than is theoretically necessary. This topic is developed further in the next section. However, it must be clearly understood that a price must be paid, and a complete evaluation of all factors and all interactions cannot be made if a confounded design is used.

The underlying basis of dividing a group of experiments into two blocks is to select the most complex interaction and to ensure that all combinations with a positive value for that interaction are in one block and all those with negative values in the other. Thus, consideration of Table 6.8 shows that in a design for a

TABLE 6.27

Two-Factor, Two-Level Factorial Design Carried Out as Two Pairs of Two Experiments

Experiment	Combinations of Pairs of Experiments		
	Arrangement 1 Loss of E (%)	Arrangement 2 Loss of E (%)	Arrangement 3 Loss of E (%)
$(-1, -1)$	10	10	10
$(+1, -1)$	25	35	35
$(-1, +1)$	40	30	40
$(+1, +1)$	55	55	45
Effect of Factor X_1	15	25	15
Effect of Factor X_2	30	20	20
Effect of Interaction X_1X_2	0	0	-10

three-factor, two-level experiment, the three-way interaction $X_1X_2X_3$ is positive in experiments $(+1, -1, -1)$, $(-1, +1, -1)$, $(-1, -1, +1)$, and $(+1, +1, +1)$ and negative in the remaining four. If the design is divided into two blocks as shown in Table 6.28, Block 1 contains all those combinations in which the three-way interaction is negative, and Block 2 all those in which it is positive. Therefore, the three-way interaction is confounded with the blocks. Any inadvertent change introduced by performing the design in two blocks is hence only confounded with the three-way interaction term. In earlier examples, the three-way interaction was often used as the "error" term in designs of this type, that is, it was assumed to have negligible significance. Hence, its confounding cannot be regarded as a major loss.

A four-factor two-level design is shown in Table 6.29. Interaction $X_1X_2X_3X_4$ is confounded with the blocks, with 1 degree of freedom, and the four three-way interactions could be pooled with 4 degrees of freedom as the error term. Such a design can be further subdivided into four blocks, as shown in Table 6.30. The first stage is division into two blocks on the basis of the four-way interaction, as described above. Then, one of the three-way interactions is chosen, and each of the two blocks is divided into two, on the basis of positive or negative signs of that interaction. Thus, Block 1 contains only negative values of the four-way interaction and negative values of the three-way interaction $X_2X_3X_4$. The blocks are confounded with

TABLE 6.28
Three-Factor, Two-Level Factorial Design Divided into Two Blocks

Block 1 ($X_1X_2X_3 = -1$)	Block 2 ($X_1X_2X_3 = +1$)
$(-1, -1, -1)$	$(+1, -1, -1)$
$(+1, +1, -1)$	$(-1, +1, -1)$
$(+1, -1, +1)$	$(-1, -1, +1)$
$(-1, +1, +1)$	$(+1, +1, +1)$

TABLE 6.29
Four-Factor, Two-Level Factorial Design Divided into Two Blocks

Block 1 ($X_1X_2X_3X_4 = -1$)	Block 2 ($X_1X_2X_3X_4 = +1$)
$(+1, -1, -1, -1)$	$(-1, -1, -1, -1)$
$(-1, +1, -1, -1)$	$(+1, +1, -1, -1)$
$(-1, -1, +1, -1)$	$(+1, -1, +1, -1)$
$(+1, +1, +1, -1)$	$(-1, +1, +1, -1)$
$(-1, -1, -1, +1)$	$(+1, -1, -1, +1)$
$(+1, +1, -1, +1)$	$(-1, +1, -1, +1)$
$(+1, -1, +1, +1)$	$(-1, -1, +1, +1)$
$(-1, +1, +1, +1)$	$(+1, +1, +1, +1)$

TABLE 6.30
Four-Factor, Two-Level Factorial Design Divided into Four Blocks

Block 1 $X_1X_2X_3X_4 = -1$ $X_2X_3X_4 = -1$	Block 2 $X_1X_2X_3X_4 = -1$ $X_2X_3X_4 = +1$	Block 3 $X_1X_2X_3X_4 = +1$ $X_2X_3X_4 = -1$	Block 4 $X_1X_2X_3X_4 = +1$ $X_2X_3X_4 = +1$
(+1, −1, −1, −1)	(−1, +1, −1, −1)	(−1, −1, −1, −1)	(+1, +1, −1, −1)
(+1, +1, +1, −1)	(−1, −1, +1, −1)	(−1, +1, +1, −1)	(+1, −1, +1, −1)
(+1, +1, −1, +1)	(−1, −1, −1, +1)	(−1, +1, −1, +1)	(+1, −1, −1, +1)
(+1, −1, +1, +1)	(−1, +1, +1, +1)	(−1, −1, +1, +1)	(+1, +1, +1, +1)

TABLE 6.31
Three-Factor, Three-Level Factorial Design Divided into Three Blocks

Block 1	Block 2	Block 3
(−1, −1, −1)	(0, −1, −1)	(+1, −1, −1)
(0, 0, −1)	(+1, 0, −1)	(−1, 0, −1)
(+1, +1, −1)	(−1, +1, −1)	(0, +1, −1)
(+1, −1, 0)	(−1, −1, 0)	(0, −1, 0)
(−1, 0, 0)	(0, 0, 0)	(+1, 0, 0)
(0, +1, 0)	(+1, +1, 0)	(−1, +1, 0)
(0, −1, +1)	(+1, −1, +1)	(−1, −1, +1)
(+1, 0, +1)	(−1, 0, +1)	(0, 0, +1)
(−1, +1, +1)	(0, +1, +1)	(+1, +1, +1)

$X_1X_2X_3X_4$, $X_2X_3X_4$, and X_1X_4. The blocks and their interactions account for 3 degrees of freedom, and the error term could be $X_1X_2X_4$, $X_1X_3X_4$, and $X_1X_2X_3X_4$, also with 3 degrees of freedom.

Blocked designs for three-level factorials are also available. The usual procedure is to arrange the experiments in blocks that are multiples of 3. Thus, a three-factor, three-level design is arranged in three blocks of nine experiments, as in Table 6.31. All main effects (X_1, X_2, and X_3) and all two-way interactions can be isolated. Examples such as this can be evaluated using the Yates method, as described earlier.

6.11 FRACTIONAL FACTORIAL DESIGNS

As stated earlier, as the number of factors in a design increases, the number of experiments needed to form a complete design can rapidly outgrow the resources available to the experimenter. If it can be safely assumed that some or all of the higher-order interactions have a negligible effect, then information on the main factors and lower-order interactions can be obtained by performing only a fraction of the total experimental design, without losing too much information.

Fractional designs are extremely useful in screening experiments where many factors are considered. Those factors that have large effects can be identified, and these can be more thoroughly investigated. Consider a three-factor, two-level design. There are eight experiments, but let us assume that only four can be carried out. Only half a full-factorial design is performed. This is conventionally represented as a 2^{3-1} design. The plus and minus signs for a three-factor, two-level design is shown in Table 6.8, and this can be divided into two blocks (Table 6.28). The three-way interaction has a value of -1 in all experiments in Block 1 and $+1$ in the experiments of Block 2.

If the experiments only in Block 2 are carried out, then the effects of the main factors and two-way interactions are calculated as follows:

$$\text{Effect of Factor } X_1 = \frac{1}{2}\left[(+1,-1,-1)-(-1,+1,-1)-(-1,-1,+1)+(+1,+1,+1)\right]$$

$$\text{Effect of Factor } X_2 = \frac{1}{2}\left[-(+1,-1,-1)+(-1,+1,-1)-(-1,-1,+1)+(+1,+1,+1)\right]$$

$$\text{Effect of Factor } X_3 = \frac{1}{2}\left[-(+1,-1,-1)-(-1,+1,-1)+(-1,-1,+1)+(+1,+1,+1)\right]$$

$$\text{Effect of Interaction } X_2X_3 = \frac{1}{2}\left[(+1,-1,-1)-(-1,+1,-1)-(-1,-1,+1)+(+1,+1,+1)\right]$$

$$\text{Effect of Interaction } X_1X_3 = \frac{1}{2}\left[-(+1,-1,-1)+(-1,+1,-1)-(-1,-1,+1)+(+1,+1,+1)\right]$$

$$\text{Effect of Interaction } X_1X_2 = \frac{1}{2}\left[-(+1,-1,-1)-(-1,+1,-1)+(-1,-1,+1)+(+1,+1,+1)\right]$$

Thus, the effect of Factor X_1 is given by an equation that is identical to that which gives the effect of Interaction X_2X_3, and so on. Consequently, it is impossible to differentiate between X_1 and X_2X_3, X_2 and X_1X_3, and X_3 and X_1X_2. The estimation of Factor X_1 is really an estimation of $(X_1+X_2X_3)$. If it is required to differentiate between the main effect and the interaction, then the other half of the design must be carried out.

Fractional factorials are available for more elaborate designs. Thus, if there are four factors — X_1, X_2, X_3, and X_4 — to be studied at two levels, but only eight experiments can be carried out from the possible 16 combinations, a suitable half-factorial fractional design (2^{4-1}) is shown in Table 6.32, which is identical to Block 2 in Table 6.29.

TABLE 6.32
Four-Factor, Two-Level Fractional Factorial Design

		Factor	
X_1	X_2	X_3	X_4
−1	−1	−1	−1
+1	+1	−1	−1
+1	−1	+1	−1
−1	+1	+1	−1
+1	−1	−1	+1
−1	+1	−1	+1
−1	−1	+1	+1
+1	+1	+1	+1

Because of the reduction in the number of experiments, considerable confounding has occurred. Thus,

Main effect X_1 is confounded with Interaction $X_2X_3X_4$
Main effect X_2 is confounded with Interaction $X_1X_3X_4$
Main effect X_3 is confounded with Interaction $X_1X_2X_4$
Main effect X_4 is confounded with Interaction $X_1X_2X_3$
Interaction X_1X_2 is confounded with Interaction X_3X_4
Interaction X_1X_3 is confounded with Interaction X_2X_4
Interaction X_2X_3 is confounded with Interaction X_1X_4

Note that there are three pairs of two-factor interactions. The results of such a design can be analyzed by Yates's method.

Further fractional designs can be devised. For example, a quarter design for a four-factor, two-level design (2^{4-2}) is based on one of the four blocks in Table 6.30. Other fractional combinations are given in Table 6.33 and Table 6.34.

TABLE 6.33
Six-Factor, Two-Level Fractional Factorial Design in Eight Experiments

			Factor		
X_1	X_2	X_3	X_4	X_5	X_6
−1	−1	−1	+1	+1	+1
+1	−1	−1	−1	−1	+1
−1	+1	−1	−1	+1	−1
+1	+1	−1	+1	−1	−1
−1	−1	+1	+1	−1	−1
+1	−1	+1	−1	+1	−1
−1	+1	+1	−1	−1	+1
+1	+1	+1	+1	+1	+1

TABLE 6.34
Seven-Factor, Two-Level Fractional Factorial Design in
Eight Experiments

			Factor			
X_1	X_2	X_3	X_4	X_5	X_6	X_7
−	−	−	+	+	+	−
+	−	−	−	−	+	+
−	+	−	−	+	−	+
+	+	−	+	−	−	−
−	−	+	+	−	−	+
+	−	+	−	+	−	−
−	+	+	−	−	+	−
+	+	+	+	+	+	+

6.12 PLACKETT–BURMAN DESIGNS

Techniques involving even more factors are available, notably those devised by Plackett and Burman.[8] They prepared two-level factorial designs for studying $N-1$ variables in N experiments, where N is a multiple of 4. If N is a power of 2, the designs are identical to those already discussed. Table 6.35 gives the combinations of −1 and +1 conditions used to construct Plackett–Burman designs for N=4, 8, 12, 16, 20, and 24. The complete designs are obtained by writing the relevant row as a column. The next column is generated by moving the elements down by one row and placing the last element in the first position. Subsequent columns are prepared in the same way. Lastly, the design is completed by adding a row of minus 1's. A Plackett–Burman design for studying 11 factors in 12 experiments is given in Table 6.36. If the number of factors is less than 11 but greater than 8, then 12 experiments must still be carried out. However, replicates can be incorporated, thus providing the error term in subsequent analysis.

Experimental designs of this type exhibit an extremely high degree of confounding. This is not surprising when one considers that a full 11-factor, 2-level design

TABLE 6.35
Combinations of −1 and +1 for the Construction of Plackett–Burman
Experimental Designs with 4, 8, 12, 16, 20, and 24 Factors

N=4	+1 +1 −1
N=8	+1 +1 +1 −1 +1 −1 −1
N=12	+1 +1 −1 +1 +1 +1 −1 −1 −1 +1 −1
N=16	+1 +1 +1 +1 −1 +1 −1 +1 +1 −1 −1 +1 −1 −1 −1
N=20	+1 +1 −1 −1 +1 +1 −1 +1 −1 +1 −1 +1 −1 −1 −1 −1 +1 +1 −1
N=24	+1 +1 +1 +1 +1 −1 +1 −1 +1 +1 −1 −1 +1 +1 −1 −1 +1 −1 +1 −1 −1 −1 −1

TABLE 6.36
Plackett–Burman Design for the Study of Eleven Factors in Twelve Experiments

					Factor					
X_1	X_2	X_3	X_4	X_5	X_6	X_7	X_8	X_9	X_{10}	X_{11}
+1	−1	+1	−1	−1	−1	+1	+1	+1	−1	+1
+1	+1	−1	+1	−1	−1	−1	+1	+1	+1	−1
−1	+1	+1	−1	+1	−1	−1	−1	+1	+1	+1
+1	−1	+1	+1	−1	+1	−1	−1	−1	+1	+1
+1	+1	−1	+1	+1	−1	+1	−1	−1	−1	+1
+1	+1	+1	−1	+1	+1	−1	+1	−1	−1	−1
−1	+1	+1	+1	−1	+1	+1	−1	+1	−1	−1
−1	−1	+1	+1	+1	−1	+1	+1	−1	+1	−1
−1	−1	−1	+1	+1	+1	−1	+1	+1	−1	+1
+1	−1	−1	−1	+1	+1	+1	−1	+1	+1	−1
−1	+1	−1	−1	−1	+1	+1	+1	−1	+1	+1
−1	−1	−1	−1	−1	−1	−1	−1	−1	−1	−1

would require 2048 (2^{11}) individual experiments, involving 11 main effects, 55 second-order interactions, and no fewer than 1981 further interactions of orders ranging from 3 to 11. Because they are so highly confounded, Plackett–Burman designs cannot be used to evaluate individual main effects and interactions. However, they are of great value in screening experiments. In these, a comparatively large number of factors may have an influence on the response, and it is hence of value to distinguish those which have an effect from those which do not. They are particularly useful in compatibility studies between active ingredients and excipients, when the two levels of the excipient are "absent" and "present."

Good examples of the use of Plackett–Burman designs in pharmaceutical systems are given by the work of Durig and Fassihi[9] and Sastry et al.[10]

6.13 CENTRAL COMPOSITE DESIGNS

Central composite designs are a progression from the factorial designs discussed earlier in this chapter and were introduced by Box and Wilson.[11] They have been widely used in response-surface modeling and optimization.

A two-factor, two-level (2^2) design as illustrated in Figure 6.4 can be developed by inclusion of a center point. A horizontal line and a vertical line are drawn through the center point, and these form the axes of a central composite design. Further experiments are positioned along the axes, at a distance α from the center point. Thus, when $X_2=0$, $X_1=\pm\alpha$, and when $X_1=0$, $X_2=\pm\alpha$, these are called axial points, and if $\alpha=1$, then the design becomes a full two-factor, three-level design, as shown in Figure 6.8.

If it is possible to go outside the original square experimental domain, the design can be extended further. For a two-factor design, the domain becomes a circle, centered on (0, 0) and passing through the factorial points (−1, −1), (+1, −1), and so on. This means that α has a value of $\sqrt{2}$, and all the axial points are therefore situated 1.414 e.u. along the axes from the center point. Essentially, the additional points are derived from a two-factor, two-level factorial design by rotating the design through 45°. Because of its shape, this design is sometimes called a star design.

A full central composite design for two factors is shown in Figure 6.10. The experimental values in coded data needed to achieve such a design are given in Table 6.37, with the center point replicated as required. There are thus five levels of each factor. The design permits a full second-order model to be investigated. The experimentation

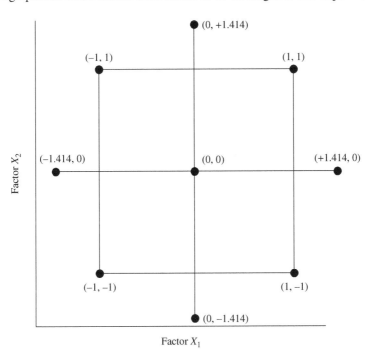

FIGURE 6.10 Central composite design for a two-factor experiment.

TABLE 6.37
Central Composite Design for a Two-Factor Experiment

	Factor X_1	Factor X_2
Center point	0	0
Factorial points	−1	−1
	−1	+1
	+1	+1
	+1	−1
"Star" points	−1.414	0
	+1.414	0
	0	−1.414
	0	+1.414

can be carried out in two blocks. The first block consists of the factorial points and a center point, and the second block the star points and replicates of the center point.

Central composite designs can also be derived for more than two factors. Thus, an experimental design for three factors, the domain of which is essentially a sphere, is shown in Figure 6.11 and Table 6.38. It consists of a three-factor, two-level design, plus six star points on the axes plus replicated points at the center of the design. The position of the "star" points is given by $\alpha=2^{N/4}$, where N is the number of factors. For a two-factor study, $\alpha=1.414$, for a three-factor study $\alpha=1.682$, and for a four-factor study $\alpha=2.000$.

If a central composite design is to be used, the designer must be confident that the values of the factors are capable of being extended outside the range of the conventional square or cubical design to encompass the star points. Few factors can have values less than zero. In the experiment shown in Figure 6.4, the catalyst concentration corresponding to −1 e.u. was zero. Therefore, values of this factor cannot be extended to $\alpha=\pm1.414$ e.u. Other factors may have minimum and maximum values. The force exerted by a tablet press cannot be less than zero, but neither can it be much greater than about 40 kN. Thus, if it is anticipated that a central composite design will be used, then the position of the center point and the magnitude of each experimental unit must be chosen with care before experimentation begins, with the factorial points located well within the limits of each variable.

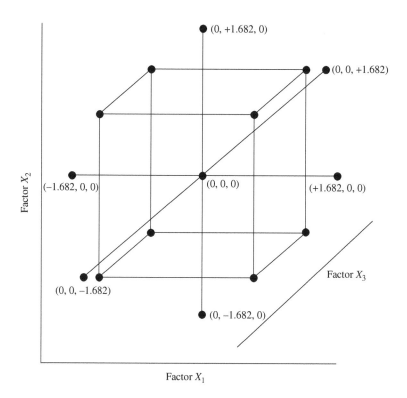

FIGURE 6.11 Central composite design for a three-factor experiment.

TABLE 6.38
Central Composite Design for a Three-Factor Experiment

	Factor X_1	Factor X_2	Factor X_3
Center point	0	0	0
Factorial points	−1	−1	−1
	+1	−1	−1
	−1	+1	−1
	+1	+1	−1
	−1	−1	+1
	+1	−1	+1
	−1	+1	+1
	+1	+1	+1
"Star" points	−1.682	0	0
	+1.682	0	0
	0	−1.682	0
	0	+1.682	0
	0	0	−1.682
	0	0	+1.682

A variation in the central composite design that reduces the number of experiments still further is the center of gravity design described by Podczeck and Wenzel.[12] The design commences with a center point. From this, at least four points are placed along the axis of each coordinate, giving a spherical design space. The extreme values of each variable will define the overall size of the experimental domain, and hence it is sensible to make these as large as possible. However, the experimental points must be located so that only realistic values of the factors are used. A center of gravity design for three factors is shown in Table 6.39. There are 13 experiments.

TABLE 6.39
Central of Gravity Design for a Three-Factor Experiment

	Factor X_1	Factor X_2	Factor X_3
Center point	0	0	0
Axial points	−2	0	0
	−1	0	0
	+1	0	0
	+2	0	0
	0	−2	0
	0	−1	0
	0	+1	0
	0	+2	0
	0	0	−2
	0	0	−1
	0	0	+1
	0	0	+2

6.14 BOX–BEHNKEN DESIGNS

In a central composite design, it is essential that every factor can be extended beyond the domain defined by the factorial points. If this is not possible, then Box–Behnken designs may be used as an alternative.[13] These use only three levels for each factor, and the domain is within the original factorial shape.

The overall structure of a three-factor Box–Behnken design is shown in Figure 6.12. The design is represented as a cube, but the experimental points are at the midpoints of the edges of the cube rather than at the corners and centers of the faces, that is, $\sqrt{2}$ or 1.414 e.u. from the center point. The values of the experimental points for this design are given in Table 6.40. Each combination of the extreme values of two of the variables is examined with the third variable having a value of zero.

The Box–Behnken design does not cover the whole of the cube of a conventional three-factor design, because the corners of the cube are not investigated. A conventional three-factor design (Figure 6.5) has a domain equal to 8 e.u.^{3}, the domain of a three-factor central composite design (Figure 6.11) has a volume of about 20 e.u.^{3}, but a three-factor Box–Behnken design has a domain volume of only 6 e.u.^{3}

A four-factor Box–Behnken design has all possible combinations of factorial design in two of the factors, with the remaining two factors taking a value of zero. There are six combinations of factors and 24 experiments plus replicated center points.

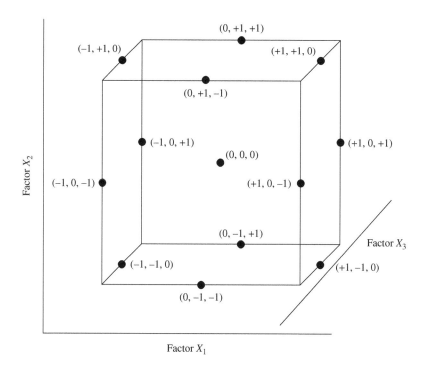

FIGURE 6.12 Box–Behnken design for a three-factor experiment.

TABLE 6.40
Box–Behnken Design for a Three-Factor Experiment

	Factor X_1	Factor X_2	Factor X_3
Center point	0	0	0
	−1	−1	0
	+1	−1	0
	−1	+1	0
	+1	+1	0
	−1	0	−1
	+1	0	−1
	−1	0	+1
	+1	0	+1
	0	−1	−1
	0	+1	−1
	0	−1	+1
	0	+1	+1

6.15 DOEHLERT DESIGNS

Essentially, a Doehlert or uniform shell design forms part of a lattice, consisting of a single point and its neighbors.[14] The Doehlert design for a two-factor experiment is a regular hexagon with a center point. There are thus seven experimental points.

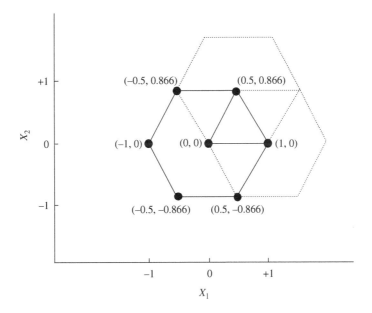

FIGURE 6.13 Doehlert design for a two-factor experiment.

**TABLE 6.41
Doehlert Design for a
Two-Factor Experiment**

X_1	X_2
0	0
1	0
0.5	0.866
-0.5	0.866
-1	0
-0.5	-0.866
0.5	-0.866

The initial stage of the design is an equilateral triangle. Starting from each point of the triangle, a separate hexagon can be constructed. The three experiments are carried out, and the "best" point is selected as the center point. This point can be a maximum or a minimum, depending on whether the "best" response has a high or low value. The initial triangle and the resulting hexagon are shown as solid lines in Figure 6.13, and the other two hexagons shown as dotted lines.

If the position of the best point is (0, 0), then the locations of the six points forming the hexagon are shown in Table 6.41. The center point of the design can be replicated as required.

Because Doehlert designs are part of a continuous network, the experimental domain can be shifted in any direction by adding experiments to one side of the

**TABLE 6.42
Doehlert Design for a Three-Factor Experiment**

X_1	X_2	X_3
0	0	0
1	0	0
0.5	0.866	0
0.5	0.289	0.816
-0.5	0.866	0
-1	0	0
-0.5	-0.866	0
0.5	-0.866	0
-0.5	0.289	0.816
0	-0.577	0.816
0.5	-0.289	-0.816
-0.5	-0.289	-0.816
0	0.577	-0.816

design and taking them away from the other, always providing that a fundamental limit, for example, a concentration of less than zero, is not breached.

A three-factor Doehlert design is constructed in a similar manner. Here, the initial design is a regular tetrahedron, the positions of the corners of which shown in the first four rows of Table 6.42. The full design of 13 experiments is given in Table 6.42, assuming that the best response is situated at (0, 0, 0).

6.16   THE EFFICIENCY OF EXPERIMENTAL DESIGNS

Because the purpose of experimental design is to improve the productivity of experimentation, it is reasonable to expect that some quantitative measures of efficiency are available. The simplest of these is termed the R-efficiency. This is the ratio of the number of coefficients in the model (P) to the number of experiments in the design (N).

Thus,

$$R_{\text{eff}} = \frac{P}{N} \le 1$$

R may alternatively be expressed as a percentage. Obviously, the smaller the number of experiments, the higher will be the value of R. However, efficiency in experimental design is not just a matter of carrying out fewer experiments. Its purpose is to carry out the experiments as effectively as possible.

A second measure of efficiency is called the D-efficiency, and its calculation involves the use of matrices (see Appendix 2). For any design within a given domain, a design is D-optimal if the determinant of the dispersion matrix, $|(X'X)^{-1}|$, is minimal. This means that the coefficients of the regression equation are estimated with maximum precision.

As described in Appendix 2, the determinant of a matrix is equal to the reciprocal of the determinant of its inverse, that is,

$$|X'X| = \frac{1}{|(X'X)^{-1}|}$$

$|X'X|$ is the information matrix, and this, in turn, can be transformed into the moment matrix M by use of equation (6.13)

$$M = N^{-1} \times X'X \tag{6.13}$$

The determinant of the moment matrix is in turn given by (6.14)

$$\left| M = \frac{|X'X|}{N^p} \right| \tag{6.14}$$

Any given experimental design (A) has a greater D-efficiency than another design (B) if $|M_A| > |M_B|$. Design efficiency is dealt with more fully by Box and Draper.[15]

FURTHER READING

A comprehensive survey of factorial designs, including a full discussion of confounding, blocked designs, and fractional designs, has been provided by Montgomery.

Montgomery, D. C., *Design and Analysis of Experiments*, 3rd ed., Wiley, New York, 1991.

The following articles describe the use of factorial techniques in the design of experiments.

GENERAL REVIEW

Bolton, S., Factorial designs in stability studies, *J. Pharm. Sci.*, 72, 362, 1983.

TWO-FACTOR, THREE-LEVEL FULL-FACTORIAL DESIGNS

Bodea, A. and Leucuta, S. E., Optimisation of propranolol hydrochloride sustained release pellets using a factorial design, *Int. J. Pharm.*, 154, 49, 1997.

Gohel, M. C. and Patel, L. D., Processing of nimesulide PEG 400 PG-PVP solid dispersions, preparation, characterisation and *in vitro* dissolution, *Drug Dev. Ind. Pharm.*, 29, 299, 2003.

Khattab, I., Menon, A., and Sakr, A., Effect of mode of incorporation of disintegrants on the characteristics of fluid-bed wet granulated tablets, *J. Pharm. Pharmacol.*, 45, 687, 1993.

Singh, B. and Ahuja, N., Development of buccoadhesive hydrophilic matrices of diltiazem HCl optimisation of bioadhesion, dissolution and diffusion parameters, *Drug Dev. Ind. Pharm.*, 28, 431, 2002.

THREE-FACTOR, TWO-LEVEL FULL-FACTORIAL DESIGNS

Chang H. C. et al., Development of a topical suspension containing three active ingredients, *Drug Dev. Ind. Pharm.*, 28, 29, 2002.

Dansereau, R., Brock, M., and Furey-Redman, N., Solubilisation of drug and excipient into a hydroxypropyl methyl cellulose based film coating as a function for the coating parameters in a 24 inch Accela-Cota, *Drug Dev. Ind. Pharm.*, 19, 793, 1993.

Iskandarani, B., Shiromani, P. K., and Claire, J. H., Scale-up feasibility in high shear mixers – detection through statistical procedures, *Drug Dev. Ind. Pharm.*, 27, 651, 2001.

Itiola, O. A. and Pilpel, N., Formulation effects on the mechanical properties of metronidazole tablets, *J. Pharm. Pharmacol.*, 43, 145, 1991.

Khanvilkar, K. H., Huand, Y., and Moore, A. D., Influence of HPMC mixture, apparent viscosity and tablet hardness on drug release using a full 2^3 factorial design, *Drug Dev. Ind. Pharm.*, 28, 641, 2002.

Li, S. F. et al., The effect of HPMC and Carbopol on the release and floating properties of a gastric floating delivery system using factorial design, *Int. J. Pharm.*, 253, 13, 2003.

THREE-FACTOR, THREE-LEVEL FULL-FACTORIAL DESIGNS

Merrku, P. and Yliruusi, J., Use of 3^3 factorial design and multilinear stepwise regression analysis in studying the fluidised bed granulation process, *Eur. J. Pharm. Biopharm.*, 39, 75, 1993.

FOUR-FACTOR, TWO-LEVEL FULL-FACTORIAL DESIGNS

Appel, L. E., Clair, J. H., and Zentner, G. M., Formulation and optimisation of a modified microporous cellulose acetate latex coating for osmotic pumps, *Pharm. Res.*, 9, 1664, 1992.

Chawla, A. et al., Production of spray dried salbutamol sulphate for use in a dry powder aerosol formulation, *Int. J. Pharm.*, 108, 233, 1994.

Dillen, K. et al., Factorial design, physicochemical characterisation and activity of ciprofloxacin-PLGA nanoparticles, *Int. J. Pharm.*, 275, 171, 2004.

Jorgensen, K. and Jacobsen, L., Factorial design used for ruggedness testing of flow through cell dissolution method by means of Weibull transformed drug release profiles, *Int. J. Pharm.*, 88, 23, 1992.

Kuentz, M. and Rothlisberger, D., Determination of the optimal amount of water in liquid fill masses for hard gelatin capsules by means of textual analysis and experimental design, *Int. J. Pharm.*, 236, 145, 2002.

Stahl, K. et al., The effect of process variables on the degradation and physical properties of spray dried insulin intended for inhalation, *Int. J. Pharm.*, 233, 227, 2002.

Vilhelmsen, T., Kristensen, J., and Schafer, T., Melt pelletisation with polyethylene glycol in a rotary processor, *Int. J. Pharm.*, 275, 141, 2004.

FOUR-FACTOR, TWO-LEVEL FRACTIONAL FACTORIAL DESIGNS

Gao, J. Z. et al., Fluid bed granulation of a poorly water soluble low density micronised drug: comparison with high shear granulation, *Int. J. Pharm.*, 237, 1, 2002.

Timmins, P. et al., Evaluation of the granulation of a hydrophilic matrix sustained release tablet, *Drug Dev. Ind. Pharm.*, 17, 531, 1991.

Worakul, N., Wongpoowarak, W., and Boonme, P., Optimisation in development of acetaminophen syrup formulations, *Drug Dev. Ind. Pharm.*, 28, 345, 2002.

FIVE-FACTOR, TWO-LEVEL FULL-FACTORIAL DESIGNS

Ku, C. C. et al., Bead manufacture by extrusion/spheronisation: statistical design for process optimisation, *Drug Dev. Ind. Pharm.*, 19, 1505, 1993.

Malinowski, H. J. and Smith, W. E., Use of factorial design to evaluate granulations prepared by spheronisation, *J. Pharm. Sci.*, 64, 1688, 1975.

FIVE-FACTOR, TWO-LEVEL FRACTIONAL FACTORIAL DESIGNS

Billon, A. et al., Development of spray dried acetaminophen microparticles using experimental design, *Int. J. Pharm.*, 203, 159, 2000.

Holm, P., Schafer, T., and Larsen, C., End point detection in a wet granulation process, *Pharm. Dev. Technol.*, 6,181, 2001.

Li, L. C. and Tu, Y. H., *In vitro* release from matrix tablets containing a silicone elastomer latex, *Drug Dev. Ind. Pharm.*, 17, 2197, 1991.

Rambali, B. et al., Using deepest regression method for optimisation of fluidised bed granulation on semi full scale, *Int. J. Pharm.*, 258, 85, 2003.

Williams, S. O. et al., Scale-up of an oil/water cream containing 40% diethylene glycol monoethyl ether, *Drug Dev. Ind. Pharm.*, 26, 71, 2000.

SIX-FACTOR, TWO-LEVEL FRACTIONAL FACTORIAL DESIGNS

Lindberg, N. O. and Jonsson, C., Granulation of lactose and starch in a recording high speed mixer: Diosna P25, *Drug Dev. Ind. Pharm.*, 11, 387, 1985.

PLACKETT–BURMAN DESIGNS

Rambali, B. et al., Using experimental design to explore the process parameters in fluid bed granulation, *Drug Dev. Ind. Pharm.*, 27, 47, 2001.

TWO-FACTOR CENTRAL COMPOSITE DESIGNS

Linden, R., et al., Response surface analysis applied to the preparation of tablets containing a high concentration of vegetable spray-dried extract, *Drug Dev. Ind. Pharm.*, 26, 441, 2000.

Martinez, S. C. et al., Acyclovir poly(D, L-lactide–co-glyceride) microspheres for intravitreal administration using a factorial design study, *Int. J. Pharm.*, 273, 45, 2003.

THREE-FACTOR CENTRAL COMPOSITE DESIGNS

Hariharen, M. and Mehdizadeh, E., The use of mixer torque rheometry to study the effect of formulation variables on properties of wet granulations, *Drug Dev. Ind. Pharm.*, 28, 253, 2002.

BOX–BEHNKEN DESIGNS

Lee, K. J. et al., Evaluation of critical formulation factors in the development of a rapidly dispersing captopril oral dosage form, *Drug Dev. Ind. Pharm.*, 29, 967, 2003.

Nazzal, S. et al., Optimisation of a self-nanoemulsified tablet dosage form of ubiquinone using response surface methodology: effect of formulation ingredients, *Int. J. Pharm.*, 240, 103, 2002.

DOEHLERT DESIGNS

Vojnovic, D. et al., Experimental research methodology applied to wet pelletisation in a high-shear mixer, *STP PHARMA Science*, 3, 130, 1993.

REFERENCES

1. Fisher, R. A., *The Design of Experiments*, Oliver & Boyd, Edinburgh, 1926.
2. Armstrong, N. A. and Cartwright, R. G., The discoloration on storage of tablets containing spray-dried lactose, *J. Pharm. Pharmacol.*, 36, 5P, 1984.
3. Yates, F., *The Design and Analysis of Factorial Experiments*, Commonwealth Agricultural Bureaux, Farnham Royal, 1959.
4. Strange, R. S., Introduction to experiment design for chemists, *J. Chem. Educ.*, 67, 113, 1990.
5. Gonzalez, A. G., Optimization of pharmaceutical formulations based on response-surface methodology, *Int. J. Pharm.*, 97, 149, 1993.
6. Plazier-Vercammen, J. A. and De Neve, R. E., Evaluation of complex formation by factorial analysis, *J. Pharm. Sci.*, 69, 1403, 1980.
7. Sanderson, I. M., Kennerley, J. W., and Parr, G. D., An evaluation of the relative importance of formulation and process variables using factorial design, *J. Pharm. Pharmacol.*, 36, 789, 1984.
8. Plackett, R. L. and Burman, J. P., The design of optimum multifactorial experiments, *Biometrika*, 33, 305, 1946.
9. Durig, T. and Fassihi, A. R., Identification of stabilising and destabilising effects of excipient–drug interactions in solid dosage form design, *Int. J. Pharm.*, 97, 161, 1993.
10. Sastry, D. V. et al., Atenolol gastrointestinal therapeutic system: 1. Screening of formulation variables, *Drug Dev. Ind. Pharm.*, 23, 157, 1997.
11. Box, G. P. E and Wilson, K. B., On the experimental attainment of optimum conditions, *J. Royal Stat. Soc. Ser. B*, 13, 1, 1951.
12. Podczeck, F. and Wenzel, U., Development of solid oral dosage forms by means of multivariate analysis: Part 3, *Pharm. Ind.*, 52, 496, 1990.
13. Box, G. E. P. and Behnken, D. W., Some new three level designs for the study of quantitative variables, *Technometrics*, 30, 95, 1960.
14. Doehlert, D. H., Uniform shell designs, *Appl. Stat.*, 19, 231, 1970.
15. Box, N. J. and Draper, N. R., Factorial designs, the $|X'X|$ criteria and some related matters, *Technometrics*, 13, 731, 1971.

7 Response-Surface Methodology

7.1 INTRODUCTION

Most experiments consist of an investigation into the relationship between two types of variables. The independent variables, or the factors, are those that are set by or under the control of the experimenter. The dependent variables, or the responses, are those that are the outcomes of the experiment. Thus, the values of the dependent variables are controlled by the values of the independent variables.

In the previous chapter, various types of experimental designs have been described. One important role of such designs is to establish the relative importance of two or more factors and also to indicate whether or not interaction occurs between these factors, thereby affecting the magnitude of the response.

Having established those factors and interactions that determine the response, the same experiments can be used for a predictive purpose, namely, estimating the response at combinations of factors that have not been studied experimentally, and this is the role of response-surface methodology. The surface can be visualized by using contour plots or three-dimensional diagrams. The prediction is carried out by deriving a mathematical model relating the response to the factors. This model is usually empirical and is based on responses to experiments that have been carried out as part of the experimental design.

Response-surface methodology permits a deeper understanding of a process or product and has many important applications. The two most important of these are in optimization and in establishing the robustness of that process or product.

The design of pharmaceutical products and processes often involves a compromise between two or more conflicting responses. For example, tablets must be strong enough to withstand the rigors of packaging, handling, and transport, yet at the same time they must comply with pharmacopoeial standards for disintegration and dissolution, both of which are adversely affected when the physical strength of the tablet is increased. Also, in virtually every process, time or cost is a limiting factor. Thus, an optimum is required, which is the best possible compromise in the given circumstances. In general terms, the optimum solution is those values of the factors which, when taken together, give the "best" values for two or more dependent variables. Optimization forms the subject of succeeding chapters.

A product or process is said to be robust if it is relatively insensitive to changes in the values of the experimental variables. It is obviously undesirable if small variations have a major impact on product or process quality. Response-surface methodology can be used to establish how robust the product is and the values of the factors that can be used to achieve maximum robustness.

The essential steps in response-surface methodology are as follows:

1. The objective of the experiment is decided upon.
2. The important factors and their interactions are identified, perhaps after a screening experiment. The levels of the factors which are to be used are also established.
3. A possible mathematical model is selected.
4. An experimental design that is appropriate to the model is chosen.
5. The experiments are carried out, and the values of the factors and the response fitted to the mathematical model.
6. The model is validated.
7. If the model does not represent the data in a satisfactory manner, then another model equation or new experimental design is selected. Stages 3, 4, 5, and 6 are repeated using models and designs of increasing complexity until a model is obtained, which is an acceptable representation of the data.
8. If required, a graphical representation of the surface is generated.

7.2 CONSTRAINTS, BOUNDARIES, AND THE EXPERIMENTAL DOMAIN

Experiments may be classified into two types: unconstrained and constrained. Consider the following quotation:

We shall defend our island, whatever the cost may be . . . we shall never surrender

Winston Churchill, June 4, 1940

This is an unconstrained situation, as the objective is to be achieved unconditionally. Consider another historic quotation:

The U.S. will land a manned spacecraft on the moon before the decade is out.

President John F. Kennedy, May, 1961

Here, there are several constraints or conditions. It is to be the United States that is to land a spacecraft on the moon, the spacecraft will be manned and a time limit has been imposed. However, the usual constraint on our actions, availability of finance, is noticeably absent!

Virtually all pharmaceutical experiments are subject to constraints or boundaries. The concentration of an ingredient cannot be less than zero, and often there is a maximum acceptable value. A piece of manufacturing equipment will pose its own constraints. Thus, a tablet press cannot exert a negative pressure, and there is also a maximum pressure above which damage to the press will ensue. These constraints form boundaries that cannot be crossed, and all experiments, irrespective of the experimental design, must be conducted within them. They must be distinguished from the experimental domain, which is the space covered by the experimental design. It is unlikely that the experiments in the design will be positioned at the constraints, and hence the domain lies within and covers a smaller space than that established by the boundaries.

7.3 RESPONSE SURFACES GENERATED FROM FIRST-ORDER MODELS

The principles of response-surface methodology will first be described by a very simple example that is then developed into a more complex situation.

The objective of the experiment is to investigate the response surface that illustrates the influence of two factors, namely, compression pressure (X_1) and disintegrant concentration (X_2), on the crushing strength of a tablet formulation (Y_1). It is proposed to use a first-order model equation (7.1)

$$Y_1 = \beta_0 + \beta_1 X_1 + \beta_2 X_2 + \varepsilon \tag{7.1}$$

The initial design is a two-factor, two-level study without replication. There are thus four experiments in the design. The chosen values of the factors are 100 MPa and 300 MPa for compression pressure and 2.5% and 7.5% for disintegrant concentration. The lower value of each factor is designated −1 and the higher value +1. The experimental design is shown in Figure 7.1, and possible constraints must now be considered.

With a linear relationship, there is no maximum or minimum value, and hence in an unconstrained situation, there is an infinite number of combinations of the two independent variables which will give a specified value of the response. However, constraints are applicable to this experiment.

1. The compression pressure X_1 cannot be less than 0 MPa. In terms of experimental units (e.u.), X_1 cannot be less than −2.
2. Likewise, the disintegrant concentration X_2 cannot be less than 0% or −2 e.u. These two constraints represent the axes of Figure 7.1.
3. X_1 cannot exceed the maximum pressure that the tablet press can safely apply. This might be 400 MPa or +2 e.u.
4. The concentration of disintegrating agent (X_2) cannot exceed a given concentration limited by the formulation. This might be 10% or +2 e.u.

These constraints are represented by the axes of Figure 7.1 and the two dotted lines drawn at +2 e.u. on each axis. However, it must be remembered that the experimental data will be generated from a much smaller domain, represented by the rectangle ABCD in Figure 7.1.

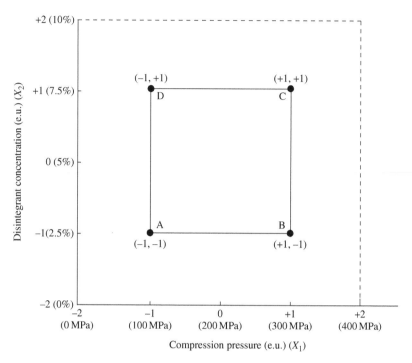

FIGURE 7.1 Two-factor, two-level factorial design to investigate the effect of compression pressure and disintegrant concentration on tablet crushing strength, showing constraints on the experimental domain.

Tablets are prepared and the measured values of their crushing strengths are shown in Table 7.1.

The relative importance of the factors and the interaction on the tablet crushing strength are now calculated as described in Chapter 6. Thus, the effect of Factor X_1, the compression pressure on crushing strength (Y_1),

TABLE 7.1

The Effect of Compression Pressure and Disintegrant Concentration on Tablet Crushing Strength, Using a Two-Factor, Two-Level Design

Experiment	Compression Pressure (MPa) (X_1)	Disintegrant Concentration (%) (X_2)	Crushing Strength (kg) (Y_1)
(−1, −1)	100	2.5	6.1
(+1, −1)	300	2.5	9.4
(−1, +1)	100	7.5	4.9
(+1, +1)	300	7.5	8.2

Note: 1 e.u. of compression pressure = 100 MPa. 1 e.u. of disintegrant concentration = 2.5%.

$$= \frac{1}{2}\{[(+1,-1)+(+1,+1)]-[(-1,-1)+(-1,+1)]\}$$

$$= \frac{1}{2}[(9.4+8.2)-(6.1+4.9)]$$

$$= 3.3$$

Similarly, the effect of Factor X_2, the disintegrant concentration, on crushing strength (Y_1)

$$= \frac{1}{2}\{[(-1,+1)+(+1,+1)]-[(-1,-1)+(+1,-1)]\}$$

$$= \frac{1}{2}[(4.9+8.2)-(6.1+9.4)]$$

$$= -1.2$$

Thus, both factors have a major influence. As might be expected, an increase in the compression pressure has a positive effect on crushing strength and an increase in the concentration of disintegrant has a negative effect.

The next stage is to fit the data from Table 7.1 into a regression equation of the form (7.1). This gives (7.2)

$$Y_1 = 7.15 + 1.65X_1 - 0.6X_2 \tag{7.2}$$

The associated statistics are:
Standard error of $b_0 = 0$
Standard error of $b_1 = 0$
Standard error of $b_2 = 0$
Standard error of $Y_1 = 0$
Coefficient of determination $(r^2) = 1.000$
$F = \infty$
Degrees of freedom $= 1$
Sum of squares of the regression equation $= 12.33$
Sum of squares of the residuals $= 0$

Because no experiment is replicated, there is no means of estimating ε or the standard errors associated with the coefficients.

The validity of the model must now be assessed. One way to achieve this is to calculate the effect of the interaction term X_1X_2 between the two factors. Interaction X_1X_2

$$= \frac{1}{2}\{[(-1,-1)+(+1,+1)]-[(+1,-1)+(-1,+1)]\}$$

$$= \frac{1}{2}[(6.1+8.2)-(9.4+4.9)]$$

$$= 0$$

Fitting the data into an equation of the form (7.3) gives a value of zero for the coefficient of the interaction term.

$$Y_1 = \beta_0 + \beta_1 X_1 + \beta_2 X_2 + \beta_{12} X_1 X_2 \qquad (7.3)$$

Another method of validation is to carry out experiments with combinations of values of factors that were not part of the original design and to compare these responses with those predicted by the model equation. The center point of the design, (0, 0) in terms of experimental units, is often chosen for this purpose because it is the point within the design space that is furthest from any of the original four experimental points. The values of both factors at the center point are zero, so the predicted value of the response at the center point is the constant term b_0, that is, 7.15 kg. If the observed value is close to this, then confidence that the model equation has validity is increased. If the observed value differs appreciably from 7.15 kg, then a more complex model may be required.

An objection to the procedure outlined above is that the experiment at the center point is carried out after the other four experiments, thus incurring the risk of confounding with respect to time. It is therefore better to include the experiment at the center point as part of the original study.

Let us assume that the experiment carried out at the center point yields tablets with a mean crushing strength of 7.3 kg; the predicted value from (7.2) is 7.15 kg. This suggests that the response surface is not planar but is curved with a peak situated somewhere within the space of the original experimental design. However, no experiment has been replicated, and hence there is no information about the underlying variability of the observations. Therefore, there is no means of knowing whether the result of 7.3 kg really represents a peak or whether it is within the range of experimental variability that can be expected. It is therefore prudent to duplicate the experiment at the center point. This will give an indication of the variability of the replicated results. If the result of this second experiment is below 7.15 kg, then the absence of a peak would be indicated, whereas another result above 7.15 kg would be further evidence of a nonplanar response surface.

Let us now assume that the two center point experiments give crushing strengths of 7.3 kg and 7.1 kg. The experimental design of six experiments, carried out simultaneously or in random order, and the results can now be summarized in Table 7.2.

The regression equation now becomes (7.4)

$$Y_1 = 7.17 + 1.65 X_1 - 0.6 X_2 \qquad (7.4)$$

The associated statistics are:
  Standard error of $b_0 = 0.036$
  Standard error of $b_1 = 0.044$
  Standard error of $b_2 = 0.044$
  Standard error of $Y_1 = 0.088$
  Coefficient of determination $(r^2) = 0.9981$
  $F = 792.6$
  Degrees of freedom $= 3$

TABLE 7.2
The Effect of Compression Pressure and Disintegrant Concentration on Tablet Crushing Strength, Using a Two-Factor, Two-Level Design with Duplicated Experiments at the Center Point

Experiment	Compression Pressure (MPa) (X_1)	Disintegrant Concentration (%) (X_2)	Observed Crushing Strength (kg) (Y_1)	Crushing Strength Predicted from (7.4) (kg)	Residual (Observed Value – Predicted Value) (kg)
(−1, −1)	100	2.5	6.1	6.12	−0.02
(+1, −1)	300	2.5	9.4	9.42	−0.02
(−1, +1)	100	7.5	4.9	4.92	−0.02
(+1, +1)	300	7.5	8.2	8.22	−0.02
(0, 0)	200	5.0	7.3	7.17	+0.13
(0, 0)	200	5.0	7.1	7.17	−0.07
			Sum of (residuals)^2 = 0.0234		

Note: Center point (0, 0): compression pressure = 200 MPa, disintegrant concentration = 5%. 1 e.u. of compression pressure = 100 MPa. 1 e.u. of disintegrant concentration = 2.5%.

Sum of squares of the regression equation = 12.33
Sum of squares of the residuals = 0.0233

From these data, it appears that a relatively simple relationship such as (7.4) provides a reasonable model. The coefficient of determination is 0.9981, indicating that over 99.8% of the variation in the responses is accounted for by the regression equation. Inclusion of an interaction term X_1X_2 in the model has no effect on the coefficient of determination.

Having established that (7.4) is a good model for the data in Table 7.2, the response surface can now be generated by substituting combinations of X_1 and X_2 into (7.4) and calculating the resultant value of the response Y_1. These are the predicted values in the fifth column of Table 7.2. The response surface is shown in Figure 7.2. The experimental domain, that is, the area covered by the experiments, is represented by a horizontal rectangle. From each corner of this rectangle, a vertical line is erected, the height of which is proportional to the size of the predicted response. The tips of these verticals are joined by straight lines, and the rectangle so formed (ABCD) is the response surface.

Because there is no interaction term, the boundaries of the response surface are straight lines. As both factors had an effect on the response, the rectangle forms a planar structure inclined in two dimensions. If one factor had had no effect on the response, then the line joining the responses at the lower and upper values of that factor would be horizontal, and hence inclination would be in only one dimension.

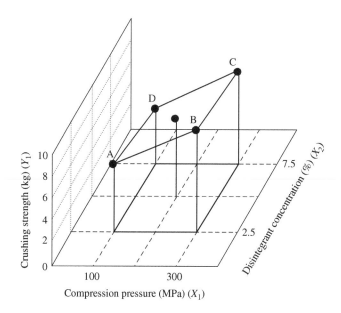

FIGURE 7.2 The response surface of tablet crushing strength (kg) as a function of compression pressure and disintegrant concentration, using responses derived from (7.4).

It is now possible to answer a question such as "if tablets of a given crushing strength (Y_1) and containing a given concentration of disintegrating agent (X_2) are required, what compression pressure (X_1) is needed to produce them?" or, in general terms, "what values of the factors are needed to give a product having specified properties?"

Equation (7.1) can be rearranged to give (7.5)

$$X_1 = \frac{\left(Y_1 - \beta_0 + \beta_2 X_2\right)}{\beta_1} \tag{7.5}$$

Y_1 and X_2 are specified, the coefficients are known, and the only unknown is X_1. For example, if the tablets are to contain 5% disintegrating agent (zero when expressed in experimental units), and should have a crushing strength of 6 kg, then (7.5) becomes (7.6)

$$X_1 = \frac{(6 - 7.17 - 0.6 \times 0)}{1.65} = \frac{-1.17}{1.65} = -0.71 \tag{7.6}$$

The required compression pressure is therefore −0.71 e.u., equivalent to 129 MPa.

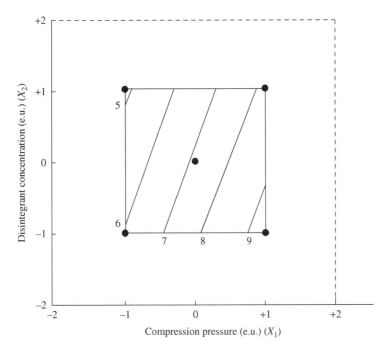

FIGURE 7.3 Contour plot of tablet crushing strength (kg) as a function of compression pressure and disintegrant concentration, using responses derived from (7.6).

By using (7.6), combinations of X_1 and X_2 can be obtained, which will give any specified value of Y_1. If these combinations are plotted, a series of straight, parallel lines is obtained (Figure 7.3). Because these lines join points of equal value of the response, they are called contour lines and Figure 7.3 is termed a contour plot.

The contour plot is a useful indicator of the robustness of the product. If the contour lines are close together, it means that a slight change in either disintegrant concentration or compression pressure would have a marked effect on tablet crushing strength. A robust product should yield a contour plot in which the lines are widely separated.

It can be seen from the contour plot that tablets with a crushing strength outside the range 5 kg to 9 kg can only be made in two very restricted areas of the original design space. This range can be extended only if the assumption is made that (7.4) applies outside the original design space.

7.4   RESPONSE SURFACES GENERATED BY MODELS OF A HIGHER ORDER

A model equation such as (7.1) can only represent a response surface that is planar. If a curved response surface is to be investigated, then a more complex equation in which factors are raised to a power of 2 or greater is needed. Such an equation is (7.7)

$$Y_1 = \beta_0 + \beta_1 X_1 + \beta_{11} X_1^2 + \beta_2 X_2 + \beta_{22} X_2^2 + \beta_{12} X_1 X_2 + \varepsilon \qquad (7.7)$$

There are six coefficients in this equation that require evaluation, and the chosen experimental design must have sufficient points to enable this to be done.

The use of higher-order designs can be explored using an experiment similar to that described earlier. In this case, the objective of the experiment is to investigate the relationship between tablet disintegration time (the response Y_2) and the factors compression pressure (X_1) and disintegrant concentration (X_2). Initially, the experimental design first described in Table 7.2 will be used, with four experiments situated at the corners of a square design space and two duplicated experiments at the center point. The constraints are the same as before. The details of the experiments and the results are shown in Table 7.3.

The relative importance of the factors and the interaction on the tablet disintegration time are calculated as described earlier. Thus the effect of Factor X_1, the compression pressure on disintegration time (Y_2)

$$= \frac{1}{2}\left\{\left[(+1,-1)+(+1,+1)\right]-\left[(-1,-1)+(-1,+1)\right]\right\}$$

$$= \frac{1}{2}\left[(1070+290)-(500+140)\right]$$

$$= 360$$

TABLE 7.3

The Effect of Compression Pressure and Disintegrant Concentration on Tablet Disintegration Time, Using a Two-Factor, Two-Level Design with Duplicated Experiments at the Center Point

Experiment	Compression Pressure (MPa) (X_1)	Disintegrant Concentration (%) (X_2)	Observed Disintegration Time (sec) (Y_2)	Disintegration Time Predicted from (7.8) (sec)	Residual (Observed Value − Predicted Value) (sec)
(−1, −1)	100	2.5	500	421.7	+78.3
(+1, −1)	300	2.5	1070	991.7	+78.3
(−1, +1)	100	7.5	140	61.7	+78.3
(+1, +1)	300	7.5	290	211.7	+78.3
(0, 0)	200	5.0	250	421.7	−171.7
(0, 0)	200	5.0	280	421.7	−141.7
			Sum of (residuals)^2 = 74084		

Note: Center point (0, 0); compression pressure = 200 MPa, disintegrant concentration = 5%. 1 e.u. compression pressure = 100 MPa. 1 e.u. of disintegrant concentration = 2.5%.

Similarly, the effect of Factor X_2, the disintegrant concentration, on disintegration time

$$= \frac{1}{2}\left\{\left[(-1,+1)+(+1,+1)\right]-\left[(-1,-1)+(+1,-1)\right]\right\}$$

$$= \frac{1}{2}\left[(140+290)-(500+1070)\right]$$

$$= -570$$

The effect of the interaction X_1X_2 on disintegration time

$$= \frac{1}{2}\left\{\left[(-1,-1)+(+1,+1)\right]-\left[(+1,-1)+(-1,+1)\right]\right\}$$

$$= \frac{1}{2}\left[(500+290)-(1070+140)\right]$$

$$= -210$$

As might be expected, increasing the compression pressure increases the tablet disintegration time and increasing the disintegrant concentration shortens it. Unlike the crushing strength data used earlier, the interaction of the factors has a relatively high negative influence.

The data are now fitted to a model equation of the form shown in (7.3), giving (7.8)

$$Y_2 = 421.7 + 180X_1 - 285X_2 - 105X_1X_2 \tag{7.8}$$

The associated statistics are:

Standard error of $b_0 = 78.57$
Standard error of $b_1 = 96.23$
Standard error of $b_2 = 96.23$
Standard error of $b_{12} = 96.23$
Standard error of $Y_2 = 192.46$
Coefficient of determination $(r^2) = 0.8706$
$F = 4.4868$
Degrees of freedom $= 2$
Sum of squares of the regression equation $= 498600$
Sum of squares of the residuals $= 74084$

Substitution of values of X_1 and X_2 into (7.8) gives the predicted values of Y_2 shown in Table 7.3.

Equation (7.8) does not fit the data particularly well. The coefficient of determination is 0.8706, showing that just over 87% of the variation among the data is explained by the regression equation. Further evidence that the equation is not a good representation of the data is obtained by examining the residuals. The residuals for the four corner points in the design are all positive, but those for the two central points are both negative. This indicates that a planar response will lie between the plane of the corner points and the position of the center point. This implies that the observed response at the center point lies in a depression, and the response surface is concave upward. Use of a model equation such as (7.7) is therefore indicated, and hence a more elaborate experimental design must be selected. Additional points must be added, and this can be done in two ways. First, a full two-factor, three-level experimental design can be used, as shown in Figure 6.8. Alternatively, a central composite design can be employed, as shown in Figure 6.10. Both of these designs involve experiments positioned at nine combinations of values of the factors.

The star or axial points in a central composite design are obtained by expanding the axes of the design to a distance equal to α e.u. from the center point. Thus, when $X_1 = 0$, $X_2 = \pm\alpha$, and when $X_2 = 0$, $X_1 = \pm\alpha$. If $\alpha = 1$, then the central composite design becomes identical to the full two-factor, three-level factorial design.

A value of $\alpha = \pm 1.414$ is usually chosen for a central composite design of two factors. This gives a circular design space with an area of 2Π square e.u. The corresponding area of the design space of a two-factor, three-level factorial design is only 4 square e.u. for the same number of experiments. Hence, for this reason, the central composite design is usually chosen. However, the choice of design is also governed by the practicability of extending the axes outside the original design space. If there is a danger of the axial points being near or even beyond the constraints, then the original four points in the design must be positioned well within these limits. The alternative is to use a three-level factorial design that does not extend the design space.

A central composite design with $\alpha = \pm 1.414$ is used to develop the experiment further. The complete design is shown in Table 7.4 and in diagrammatic form in Figure 7.4.

There are now four experiments carried out at the center point. The initial experimental design involved four experiments at the corners of the initial design space plus two experiments at the center point. It was only after these six experiments had been evaluated that a decision was taken to use a central composite design, that is, the whole of the central composite design was not carried out at the same time or in random order. Therefore, two further center point experiments are carried out in association with the star points to ensure that the overall design has not been confounded by being carried out in two blocks.

Regression of the tablet disintegration time results in Table 7.4 gives (7.9), together with associated statistical information

$$Y_2 = 261.3 + 174.86X_1 + 179.06X_1{}^2 - 230.03X_2 + 100.31X_2{}^2 - 105X_1X_2 \quad (7.9)$$

The associated statistics are
  Standard error of $b_0 = 39.90$
  Standard error of $b_1 = 28.22$
  Standard error of $b_{11} = 31.55$
  Standard error of $b_2 = 28.22$
  Standard error of $b_{22} = 31.55$
  Standard error of $b_{12} = 39.90$
  Standard error of $Y_2 = 79.81$
  Coefficient of determination $(r^2) = 0.9611$

TABLE 7.4
The Effect of Compression Pressure and Disintegrant Concentration on Tablet Disintegration Time, Using a Central Composite Design

Experiment	Compression Pressure (MPa) (X_1)	Disintegrant Concentration (%) (X_2)	Observed Disintegration Time (sec) (Y_2)	Disintegration Time Predicted from (7.9) (sec)	Residual (Observed Value – Predicted Value) (sec)
(−1, −1)	100	2.5	500	490.8	+9.2
(+1, −1)	300	2.5	1070	1050.5	+19.5
(−1, +1)	100	7.5	140	240.7	−100.7
(+1, +1)	300	7.5	290	380.5	−90.5
(−1.414, 0)	59	5.0	420	373.2	+46.8
(+1.414, 0)	341	5.0	900	865.6	+34.4
(0, −1.414)	200	4.0	750	787.1	−37.1
(0, +1.414)	200	6.0	255	136.6	+118.4
(0, 0)	200	5.0	250	261.3	−11.3
(0, 0)	200	5.0	280	261.3	+18.7
(0, 0)	200	5.0	250	261.3	−11.3
(0, 0)	200	5.0	265	261.3	+3.7

Sum of
$(\text{residuals})^2 = 38183$

Note: Center point (0, 0); compression pressure = 200 MPa, disintegrant concentration = 5%. 1 e.u. of compression pressure = 100 MPa. 1 e.u. of disintegrant concentration = 2.5%.

$F = 29.67$
Degrees of freedom = 6
Sum of squares of the regression equation = 944861
Sum of squares of the residuals = 38183

The residuals are also shown in Table 7.4. The quadratic equation is a much better fit to the data and thus is a more appropriate model than (7.8).

A contour plot of the disintegration time data can now be constructed. Equation (7.7) can be rearranged to give (7.10), which is a quadratic equation in terms of X_1.

$$\beta_{11}X_1^2 + (\beta_1 + \beta_{12}X_2)X_1 + (\beta_2X_2 + \beta_{22}X_2^2 + \beta_0 - Y_2) = 0 \qquad (7.10)$$

There are two solutions to a quadratic equation, given by (7.11)

$$X = \frac{-b \pm \sqrt{b^2 - 4ac}}{2a} \qquad (7.11)$$

where
a, b, and c = the coefficients of X raised to the powers 2, 1, and 0, respectively.

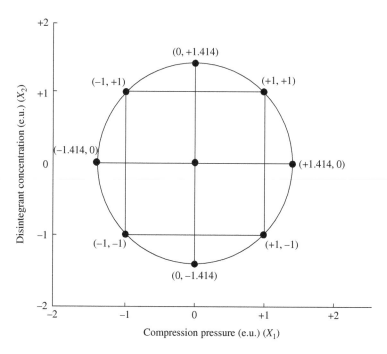

FIGURE 7.4 Central composite design to investigate the effect of compression pressure and disintegrant concentration on tablet disintegration time.

For (7.10), $a = \beta_{11}$, $b = (\beta_1 + \beta_{12}X_2)$, and $c = (\beta_2 X_2 + \beta_{22} X_2^2 + \beta_0 - Y_2)$. Therefore, the two solutions to (7.10) are given by (7.12) and (7.13)

$$X_1 = \frac{-(\beta_1 + \beta_{12}X_2) - \sqrt{(\beta_1 + \beta_{12}X_2)^2 - 4\beta_{11}(\beta_2 X_2 + \beta_{22} X_2^2 + \beta_0 - Y_2)}}{2\beta_{11}} \qquad (7.12)$$

$$X_1 = \frac{-(\beta_1 + \beta_{12}X_2) + \sqrt{(\beta_1 + \beta_{12}X_2)^2 - 4\beta_{11}(\beta_2 X_2 + \beta_{22} X_2^2 + \beta_0 - Y_2)}}{2\beta_{11}} \qquad (7.13)$$

The values of the coefficients b_0, b_1, b_{11}, b_2, b_{22}, and b_{12} have been obtained by multiple regression, and hence for specified values of X_2 and Y_2, the required value of X_1 can be calculated. The use of a computer spreadsheet is invaluable at this point.

Figure 7.5 shows a contour plot of disintegration time as a function of compression pressure and disintegrant concentration. The same constraints apply as before. The contours are curved and are located within a circular experimental domain, the radius of which is 1.414 e.u.

Cross sections through the response surface are illustrated in Figure 7.6. The first row of diagrams shows the disintegration times when X_1, the compression pressure, equals −1 e.u., 0 e.u., and +1 e.u., respectively. The second row illustrates the analogous situation when X_2, the concentration of disintegrant, equals −1 e.u., 0 e.u., and +1 e.u. The shallower the curve, the more robust the formulation.

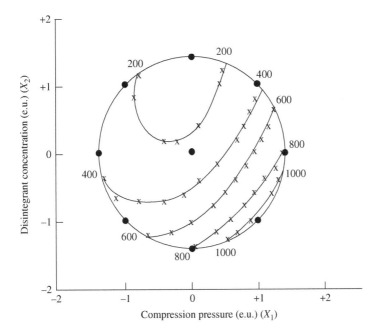

FIGURE 7.5 Contour plot of tablet disintegration time (s) as a function of compression pressure and disintegrant concentration, using responses derived from (7.9).

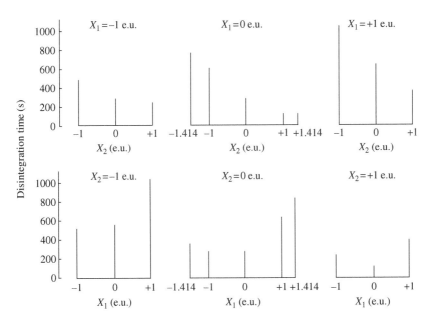

FIGURE 7.6 Cross sections through the response surface of tablet disintegration time (s) as a function of compression pressure and disintegrant concentration, using responses derived from (7.9).

Models of a higher order than 2 are rarely used in response-surface methodology. Considerably more experiments are needed, and also the predictive power of such models is often low, especially if extrapolation is involved.

7.5 RESPONSE-SURFACE METHODOLOGY WITH THREE OR MORE FACTORS

As the number of factors increases, so does the complexity of the model equation, and because of this, a more elaborate experimental design is needed. Thus, if three factors are studied at two levels, then the relevant equation would be (7.14), with three two-way interactions and one three-way interaction.

$$Y = \beta_0 + \beta_1 X_1 + \beta_2 X_2 + \beta_3 X_3 + \beta_{12} X_1 X_2 + \beta_{13} X_1 X_3 + \beta_{23} X_2 X_3 + \beta_{123} X_1 X_2 X_3 + \varepsilon \quad (7.14)$$

A three-factor, two-level full-factorial design as shown in Figure 6.8 could be used to study this. However, this design has only eight experiments and hence is a saturated design as (7.14) has eight coefficients. There is no possibility of validation, and because no experiment is replicated, no indication of the error term ε can be made.

If a quadratic relationship is sought, then (7.15) would be used.

$$\begin{aligned}
Y = \beta_0 + \beta_1 X_1 + \beta_{11} X_1^2 + \beta_2 X_2 + \beta_{22} X_2^2 + \beta_3 X_3 + \beta_{33} X_3^2 \\
+ \beta_{12} X_1 X_2 + \beta_{13} X_1 X_3 + \beta_{23} X_2 X_3 + \beta_{123} X_1 X_2 X_3 + \varepsilon
\end{aligned} \quad (7.15)$$

This equation has 11 coefficients, so a design of at least 11 experiments plus replication at the center point is needed. These extra experiments could be supplied in many ways. First, a full three-factor, three-level design could be used. This has 27 experiments plus replication at the center point. An alternative is a three-factor Box–Behnken design (Figure 6.12) of 13 points plus center point replication. Such a design was employed by Tattawasart and Armstrong[1] in a study of the properties of lactose plugs used as fills of hard-shell capsules. The three factors were lubricant concentration (X_1), applied pressure (X_2), and dosator piston height (X_3), and six properties of the plug formed the responses. Details of the design are given in Table 7.5. The Box–Behnken design was particularly useful in this case because of the limited range of the dosator piston height setting. Extension of the design space to encompass a central composite design was not feasible. Also, as piston movement could only be changed in increments of 1.0 mm, the "center point" of the design had a value of +0.14 e.u. for factor X_3 rather than zero.

A central composite design may also be appropriate for experiments with three or more factors. By using a model of this type, Pourkavoos and Peck[2] studied the effect of three factors involved in tablet film coating conditions. The factors were inlet air temperature (X_1), coating pan rotational speed (X_2), and coating solution

TABLE 7.5

Three-Factor Box–Behnken Design to Investigate Factors Involved in Lactose Plug Formation

Variable	−1	0 (+0.14 for X_3)	+1
X_1 (%w/w)	0.5	1.0	1.5
X_2 (MPa)	4	6	8
X_3 (mm)	8	12	15

X_1	X_2	X_3
−1	−1	+0.14
−1	0	−1
−1	+1	+0.14
−1	0	+1
0	−1	−1
0	+1	−1
0	+1	+1
0	−1	+1
+1	−1	+0.14
+1	0	−1
+1	+1	+0.14
+1	0	+1
0	0	+0.14
0	0	+0.14

Source: Tattawasart and Armstrong[1]

spray rate (X_3). The central composite design is shown in Table 7.6, which in turn was divided into three orthogonal blocks. All experiments in a block were carried out on the same day in a randomized order. Five responses were measured.

For four independent variables studied at two levels, the model equation becomes (7.16)

TABLE 7.6

Three-Factor Central Composite Design to Investigate Factors Involved in Tablet Film Coating

Variable	−1.633	−1	0	+1	+1.633
X_1 (°C)	50	55	62.5	70	75
X_2 (rpm)	10	11	12.5	14	15
X_3 (g·min^{-1})	75	79	85	91	94

Source: Pourkavoos and Peck[2]

$$Y = \beta_0 + \beta_1 X_1 + \beta_2 X_2 + \beta_3 X_3 + \beta_4 X_4 + \beta_{12} X_1 X_2 + \beta_{13} X_1 X_3 \qquad (7.16)$$
$$+ \beta_{14} X_1 X_4 + \beta_{23} X_2 X_3 + \beta_{24} X_2 X_4 + \beta_{34} X_3 X_4 + \varepsilon$$

This has 11 coefficients, and hence there must be a minimum of 11 experiments in the design. A complete four-factor, two-level design consists of 16 experiments, but some points must be replicated so that the error term can be estimated.

An example using this model is the work of Iskandarani et al.[3] These workers selected four independent variables involved in capsule and tablet formulation. These were the quantity of granulating agent (X_1), quantity of lubricant (X_2), quantity of granulating solution (X_3), and quantity of disintegrating agent (X_4) in each unit. Six capsule and tablet properties were measured as the responses.

By choosing to ignore all interactions involving X_4, Iskandarani et al. were able to reduce (7.16) to (7.17), with a consequent simplification of the experimental design. They were able to use a fractional four-factor design with duplicated experiments at the center point. Their design is shown in Table 7.7.

$$Y = \beta_0 + \beta_1 X_1 + \beta_2 X_2 + \beta_3 X_3 + \beta_4 X_4 + \beta_{12} X_1 X_2 + \beta_{13} X_1 X_3 + \beta_{23} X_2 X_3 + \varepsilon \qquad (7.17)$$

If a quadratic model with four factors is to be used, the model equation becomes (7.18).

TABLE 7.7
Four-Factor Fractional Factorial Design to Investigate Factors Involved in Capsule and Tablet Formulation

Variable	−1	0	+1
X_1 (mg)	3.00	5.00	13.00
X_2 (mg)	0.85	1.70	2.50
X_3 (mg)	23.20	31.00	38.60
X_4 (mg)	2.00	5.00	8.00

X_1	X_2	X_3	X_4
1	1	1	1
1	1	−1	−1
1	−1	1	−1
1	−1	−1	1
−1	1	1	1
−1	1	−1	1
−1	−1	1	1
−1	−1	−1	−1
0	0	0	0
0	0	0	0

Source: Iskandarani et al.[3]

$$Y = \beta_0 + \beta_1 X_1 + \beta_{11} X_1^2 + \beta_2 X_2 + \beta_{22} X_2^2 + \beta_3 X_3 + \beta_{33} X_3^2 + \beta_4 X_4 + \beta_{44} X_4^2 \quad (7.18)$$
$$+ \beta_{12} X_1 X_2 + \beta_{13} X_1 X_3 + \beta_{14} X_1 X_4 + \beta_{23} X_2 X_3 + \beta_{24} X_2 X_4 + \beta_{34} X_3 X_4 + \varepsilon$$

This equation has 15 coefficients.

One of the earliest applications of response-surface methodology to pharmaceutical systems was that published by Schwartz et al.[4] These workers looked at five independent variables relevant to a tablet formulation. These were diluent composition (X_1), compression pressure (X_2), disintegrant content (X_3), granulating agent content (X_4), and lubricant content (X_5). For each formulation, eight responses were measured. These were disintegration time, hardness, dissolution rate, friability, weight, thickness, porosity, and mean pore diameter. Each response was fitted into an equation containing all independent variables up to the power 2 and all two-way interactions (7.19)

$$Y = \beta_0 + \beta_i X_1 + \beta_{11} X_1^2 + \cdots + \beta_5 X_5 + \beta_{55} X_5^2 + \beta_{12} X_1 X_2 + \cdots + \beta_{45} X_4 X_5 \quad (7.19)$$

This equation contains 21 unknown coefficients, and hence an experimental design had to be chosen which would provide sufficient data points for these to be calculated.

The design used by Schwartz et al., comprising 27 experiments, is shown in Table 7.8. Experimental units are used throughout, and the experiments are numbered 1–27 for convenience. The first 16 experiments represent a half-factorial design for five factors at two levels. A full-factorial design would require 32 experiments even with no replication, and hence the reduction to 16 leads to some confounding. However, none of the two-way interactions are confounded with main effects or with each other, but three-way interactions are considerably confounded, for example, $X_1 X_2 X_3$ with $X_4 X_5$.

The remainder of the 27 experiments are needed to provide sufficient experimental points and also to achieve symmetry. Thus, for each factor, three additional levels were selected. Zero represents the midpoint of each factor in the design, and −1.547 and +1.547 are the extreme values of each variable. Experiment 27 represents the midpoint of the whole design, with all five factors set to 0 e.u.

In this study, factor X_5 is lubricant content in milligrams, and one experimental unit represents 0.5 mg of lubricant. Therefore, the five levels of lubricant—which in experimental units are −1.547, −1, 0, +1, and +1.547—are, when expressed in physical units, 0.2, 0.5, 1.0, 1.5, and 1.8 mg, respectively. A full translation of experimental units into physical units is shown in Table 7.8.

The reason for selecting a value of $\alpha = \pm 1.547$ as the extra values of the factors is not given in the original paper by Schwartz et al. A central composite design of five factors would have its star points at a distance α from the center point of $2^{5/4} = 2.378$ e.u. If this value of α were to be applied to the design used by Schwartz et al., then some of the values of the factors would be outside the constraints of the experimental design, with, for example, negative values for X_2, X_4, and X_5.

It is impossible to diagrammatically represent a contour plot or the response surface with more than two factors in the design. However, this can be obtained if

TABLE 7.8
Five Factor Fractional Factorial Design to Investigate Factors Involved in Tablet Formulation

Designation	Description	Factor 1 e.u. Equivalent	−1.547	−1	0	+1	+1.547
X_1	Diluent	10 mg	24.5	30	40	30	55.50
X_2	Compression pressure	0.5 ton	0.25	0.5	1	1.5	1.75
X_3	Disintegrant	1 mg	2.5	3	4	5	5.5
X_4	Granulating agent	0.5 mg	0.2	0.5	1	1.5	1.8
X_5	Lubricant	0.5 mg	0.2	0.5	1	1.5	1.8

Experiment	Factor Level (In Experimental Unit)				
	X_1	X_2	X_3	X_4	X_5
1	−1	−1	−1	−1	+1
2	+1	−1	−1	−1	−1
3	−1	+1	−1	−1	−1
4	+1	+1	−1	−1	+1
5	−1	−1	+1	−1	−1
6	+1	−1	+1	−1	+1
7	−1	+1	+1	−1	+1
8	+1	+1	+1	−1	−1
9	−1	−1	−1	+1	−1
10	+1	−1	−1	+1	+1
11	−1	+1	−1	+1	+1
12	+1	+1	−1	+1	−1
13	−1	−1	+1	+1	+1
14	+1	−1	+1	+1	−1
15	−1	+1	+1	+1	−1
16	+1	+1	+1	+1	+1
17	−1.547	0	0	0	0
18	+1.547	0	0	0	0
19	0	−1.547	0	0	0
20	0	+1.547	0	0	0
21	0	0	−1.547	0	0
22	0	0	+1.547	0	0
23	0	0	0	−1.547	0
24	0	0	0	+1.547	0
25	0	0	0	0	−1.547
26	0	0	0	0	+1.547
27	0	0	0	0	0

Source: Schwartz et al.[4]

all but two of the factors are kept constant. Thus, for a three-factor experiment in the form of a cube, a contour plot is obtained by taking a "slice" through the design, for example, by joining the points $(0, -1, -1)$, $(0, +1, -1)$, $(0, +1, +1)$, and $(0, -1, +1)$. Thus, the variation in factors X_2 and X_3 can be evaluated, factor X_1 being constant at a value of 0 e.u.

FURTHER READING

Myers, R. H. and Montgomery, D. C., *Response Surface Methodology: Process and Product Optimisation using Designed Experiments*, 2nd ed., Wiley, New York, 2002.

Bodea, A. and Leucuta, S. E., Optimisation of propranolol hydrochloride sustained release pellets using a factorial design, *Int. J. Pharm.*, 154, 49, 1997.

Chang, H. C. et al., Development of a topical suspension containing 3 active ingredients, *Drug Dev. Ind. Pharm.*, 28, 29, 2002.

Gohel, M. C., Patel, M. M., and Amin, A. F., Development of modified release diltiazem HCl tablets using composite index to identify optimal formulation, *Drug Dev. Ind. Pharm.*, 29, 565, 2003.

Iskandarani, B., Shiromani, P. K., and Claire, J. H., Scale-up feasibility in high shear mixers – detection through statistical procedures, *Drug Dev. Ind. Pharm.*, 27, 651, 2001.

Kiekens, F. et al., Influence of the punch diameter and curvature on the yield pressure of microcrystalline cellulose compacts during Heckel analysis, *Eur. J. Pharm. Sci.*, 22, 117, 2004.

Kuentz, M. and Rothlisberger, D., Determination of the optimal amount of water in liquid fill masses for hard gelatin capsules by means of textual analysis and experimental design, *Int. J. Pharm.*, 236, 145, 2002.

Levina, M. and Rubinstein, M. H., Effect of ultrasonic vibration on compaction characteristics of ibuprofen, *Drug Dev. Ind. Pharm.*, 28, 495, 2002.

Lewis, G. A. and Chariot, M., Non-classical experimental designs in pharmaceutical formulation, *Drug Dev. Ind. Pharm.*, 17, 1551, 1991.

Linden, R. et al., Response surface analysis applied to the preparation of tablets containing a high concentration of vegetable spray-dried extract, *Drug Dev. Ind. Pharm.*, 26, 441, 2000.

Magee, G. A. et al., Bile salt/lecithin mixed micelles optimised for the solubilisation of a poorly soluble steroid molecule using statistical experimental design, *Drug Dev. Ind. Pharm.*, 29, 441, 2003.

Rambali, B. et al., Itraconazole formulation studies of the melt extrusion process with mixture design, *Drug Dev. Ind. Pharm.*, 29, 641, 2003.

Singh, B. and Ahuja, N., Development of buccoadhesive hydrophilic matrices of diltiazem HCl: optimisation of bioadhesion, dissolution and diffusion parameters, *Drug Dev. Ind. Pharm.*, 28, 431, 2002.

Vilhelmsen, T., Kristensen, J., and Schafer, T., Melt pelletisation with polyethylene glycol in a rotary processor, *Int. J. Pharm.*, 275, 141, 2004.

Ye Huang, K. H. et al., Effects of manufacturing process variables on the *in vitro* dissolution characteristics of extended release tablets formulated with HPMC, *Drug Dev. Ind. Pharm.*, 29, 79, 2003.

REFERENCES

1. Tattawasart, A. and Armstrong, N. A., The formation of lactose plugs for hard shell capsule fills, *Pharm. Dev. Technol.*, 2, 335, 1997.
2. Pourkavoos, N. and Peck, G. E., Effect of aqueous film coating conditions on water removal efficiency and physical properties of coated tablet cores containing superdisintegrants, *Drug Dev. Ind. Pharm.*, 20, 1535, 1994.

3. Iskandarani, B., Clair, J. H., Patel, P., Shiromani, P. K., and Dempski, R. E., Simultaneous optimization of capsule and tablet formulation using response surface methodology, *Drug Dev. Ind. Pharm.*, 19, 2089, 1993.
4. Schwartz, J. B., Flamholz, J. R., and Press, R. H., Computer optimisation of pharmaceutical formulations. 1. General procedure. *J. Pharm. Sci.*, 62, 1168, 1973.

8 Model-Dependent Optimization

8.1 INTRODUCTION

Most experiments consist of an investigation into the relationship between the independent variable(s) and one or more dependent variables.

In some cases, there is only one dependent variable of interest. The values of the independent variables are chosen so that the process is maximized or minimized, or to obtain some predetermined target. However, sometimes, there may be two or more responses, both of which must be considered. Unless the two responses are highly correlated, it is unlikely that the values of the independent variables needed to achieve the maximum value of one response will be the same as those needed for the maximum value of a second response. Hence, the most favorable solution must be sought bearing in mind the values of both responses. This is termed the optimum solution, and the process is known as optimization or multicriteria decision-making.

There are many strategies available that can be used to determine the position of the optimum response, and these can conveniently be divided into two groups: sequential methods and simultaneous methods. The former commences with the performance of a small number of experiments. The results of these are considered, and a further small number of experiments is carried out, followed by further consideration. The process is repeated until the optimum solution is reached. This is analogous to climbing a hill in poor visibility. By proceeding ever upward in small steps, a summit can eventually be reached. Sequential or model-independent methods are discussed in Chapter 9.

Simultaneous methods are the alternative to sequential methods. Here, a complete set of experiments is performed, after which mathematical modeling takes place, usually by employing regression techniques. This enables the position of the optimum to be calculated. It can be likened to finding the summit of a hill by preparing a contour map, joining together points of equal altitude. The experimental designs are usually some form of factorial design, as described in Chapter 6.

Irrespective of which method is used, some idea of the relationships between the responses and the independent variables is obtained. This is termed the response surface and is described in Chapter 7. Many techniques for studying the response surface have been developed, and the subject has been comprehensively reviewed by Myers and Montgomery.[1]

8.2 MODEL-DEPENDENT OPTIMIZATION

In Chapter 7, an example based on tablet formulation was used to illustrate response-surface methodology. The two independent variables were compression pressure and disintegrant concentration, and the two responses were tablet crushing strength and disintegration time. This is a classical case of optimization where a compromise is needed. Increased compression pressure will raise crushing strength, but it will prolong disintegration time. Conversely, increasing the disintegrant concentration shortens the disintegration time of the tablets but also weakens them.

The objective of the experiment now becomes to find the values of the independent variables that will give the optimum tablet properties, bearing in mind both crushing strength and disintegration time. The constraints on the experimental design are as before, namely:

1. The compression pressure X_1 cannot be less than 0 MPa. In terms of experimental units (e.u.), X_1 cannot be less than −2.
2. Likewise, the disintegrant concentration X_2 cannot be less than 0% or −2 e.u. These two constraints represent the axes of Figure 8.1 and Figure 8.2.
3. X_1 cannot exceed the maximum pressure that the tablet press can safely apply. This might be 400 MPa or +2 e.u.
4. The concentration of disintegrating agent (X_2) cannot exceed a given concentration limited by the formulation. This might be 10% or +2 e.u. In addition, there is a fifth constraint that is
5. Y_2 cannot be greater than 900 seconds (the European Pharmacopoeial limit for tablet disintegration time).

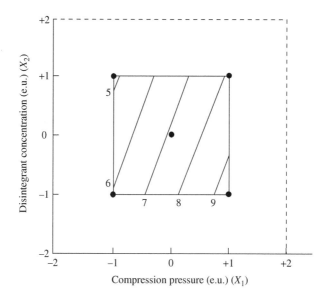

FIGURE 8.1 Contour plot of tablet crushing strength derived from a two-factor, central composite design with replicated central points.

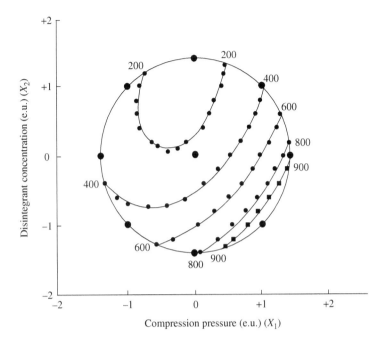

FIGURE 8.2 Contour plot of tablet disintegration time derived from a two-factor central composite design with replicated central points.

Because it is anticipated that the interaction term X_1X_2 and the terms in which X_1 and X_2 are raised to the power 2 might be significant, a central composite design is chosen. The experimental conditions and the responses are shown in Table 8.1.

The first stage in the optimization procedure is to carry out multiple regression analysis to derive model equations, as described in Chapters 4 and 7. Three possible models will be investigated, these having the forms shown in (8.1), (8.2), and (8.3).

$$Y_n = \beta_0 + \beta_1 X_1 + \beta_2 X_2 + e \tag{8.1}$$

$$Y_n = \beta_0 + \beta_1 X_1 + \beta_2 X_2 + \beta_{12} X_1 X_2 + e \tag{8.2}$$

$$Y_n = \beta_0 + \beta_1 X_1 + \beta_{11} X_1^2 + \beta_2 X_2 + \beta_{22} X_2^2 + \beta_{12} X_1 X_2 + e \tag{8.3}$$

Considering the crushing strength data first, fitting them into (8.1) yields (8.4)

$$Y_1 = 7.17 + 1.65 X_1 - 0.60 X_2 \tag{8.4}$$

TABLE 8.1
The Effect of Compression Pressure (X_1) and Disintegrant Concentration (X_2) on Tablet Crushing Strength (Y_1) and Tablet Disintegration Time (Y_2), Using a Central Composite Design with Replicated Experiments at the Center Point

Experiment	Compression Pressure (MPa) (X_1)	Disintegrant Concentration (%) (X_2)	Tablet Crushing Strength (kg) (Y_1)	Disintegration Time (sec) (Y_2)
(−1, −1)	100	2.5	6.1	500
(+1, −1)	300	2.5	9.4	1070
(−1, +1)	100	7.5	4.9	140
(+1, +1)	300	7.5	8.2	790
(−1.414, 0)	60	5.0	4.8	420
(+1.414, 0)	340	5.0	9.5	900
(0, −1.414)	200	1.5	8.0	750
(0, +1.414)	200	8.5	6.3	255
(0, 0)	200	5.0	7.3	250
(0, 0)	200	5.0	7.1	280
(0, 0)	200	5.0	7.2	250
(0, 0)	200	5.0	7.2	265

Note: Center point (0, 0): compression pressure = 200 MPa, disintegrant concentration = 5%. 1 e.u. of compression pressure = 100 MPa. 1 e.u. of disintegrant concentration = 2.5%.

with a coefficient of determination (r^2) of 0.9989. Use of (8.2) or (8.3) has no effect on the coefficient of determination and gives negligible values for β_{11}, β_{22}, and β_{12}. Hence, (8.4) can be regarded as a satisfactory representation of the relationship of Y_1 with X_1 and X_2.

The disintegration time data are treated similarly and yields the three regression equations (8.5), (8.6), and (8.7). The coefficient of determination is given in each case.

$$Y_2 = 447.50 + 174.87X_1 - 230.03X_2 \quad r^2 = 0.6793 \tag{8.5}$$

$$Y_2 = 447.50 + 174.87X_1 - 230.03X_2 - 105.00X_1X_2 \quad r^2 = 0.7242 \tag{8.6}$$

$$Y_2 = 261.25 + 174.87X_1 + 179.06X_1{}^2 - 230.03X_2 \tag{8.7}$$
$$+ 100.31X_2{}^2 - 105.00X_1X_2 \quad r^2 = 0.9611$$

It is useful at this stage to check the signs of the coefficients. An increase in compression pressure should prolong disintegration time, and hence a positive sign would be expected for coefficient β_1. Similarly, a negative sign for the coefficient β_2 would reflect the expected outcome of the disintegration time, being reduced in the presence of a greater amount of disintegrating agent.

Equation (8.7) is a much better representation of the relationship of disintegration time with compression pressure and disintegrant concentration than are (8.5) and (8.6).

Equation (8.4) can be rearranged to form (8.8), as described in the previous chapter, so that for given values of X_2, the required value of X_1 is obtained to give a predetermined value of the response.

$$X_1 = \frac{(Y_1 - 7.17 + 0.60X_2)}{1.65} \tag{8.8}$$

A contour plot joining points of equal values of crushing strength can now be constructed from (8.8), giving Figure 8.1, which is a series of straight lines.

The solution for (8.7), a quadratic equation, is more complex. There are two possible solutions: (8.9) and (8.10)

$$X_1 = \frac{-(\beta_1 + \beta_{12}X_2) - \sqrt{(\beta_1 + \beta_{12}X_2)^2 - 4\beta_{11}(\beta_2 X_2 + \beta_{22}X_2^2 + \beta_0 - Y_2)}}{2\beta_{11}} \tag{8.9}$$

$$X_1 = \frac{-(\beta_1 + \beta_{12}X_2) + \sqrt{(\beta_1 + \beta_{12}X_2)^2 - 4\beta_{11}(\beta_2 X_2 + \beta_{22}X_2^2 + \beta_0 - Y_2)}}{2\beta_{11}} \tag{8.10}$$

The corresponding contour plot, a series of curves, is shown in Figure 8.2.

8.2.1 EXTENSION OF THE DESIGN SPACE

There is no certainty that the combinations of factors chosen for the original experimental design will encompass an area in which the optimum of the responses is situated. Figure 8.1 and Figure 8.2, considered individually, will give indications of the range of responses which can be achieved within the constraints. The possibility must now be considered that the selected combinations of experimental conditions may not give tablets possessing the required properties. For example, imagine that there is a constraint to the effect that tablets must have a minimum crushing strength of 10 kg. This is greater than the strength of any of the tablets reported in Table 8.1. It is tempting to extrapolate and calculate combinations of conditions that will give tablets of the required strength. However, the inherent hazards of this approach must be borne in mind, because there is no evidence that (8.4) applies outside the experimental range within which it was derived.

A more satisfactory procedure is to extend the study using one or more additional factorial designs. An example of this is shown in Figure 8.3. A line is drawn passing through the coordinates of the center point of the study and perpendicular to the series of contours. This is known as the path of steepest ascent. The point of intersection between this line and the "box" of the original design then forms the center point of the next factorial design. This intersection is at point A, the coordinates of which are +1 e.u. of compression pressure and −0.4 e.u. of disintegrant concentration, equivalent to a pressure of 300 MPa and 4.0% disintegrant. Thus, a suitable second two-factor, two-level design would be to use the combination of values of factors shown in Table 8.2. The experiments are performed, linear regression

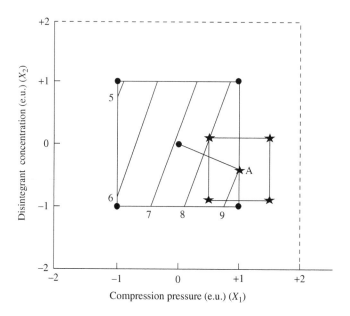

FIGURE 8.3 Procedure for determining the position of a second factorial design, using the path of steepest ascent method.

TABLE 8.2
Combination of Factors for a Second Two-Factor, Two-Level Design after Use of the Path of Steepest Ascent Technique on the Data Presented in Table 8.1

Experiment	Compression Pressure (MPa) (X_1)	Disintegrant Concentration (%) (X_2)
(+0.5, −0.9)	250	2.75
(+1.5, −0.9)	350	2.75
(+0.5, +0.1)	250	5.25
(+1.5, +0.1)	350	5.25
(+1, −0.4)	300	4.00
(+1, −0.4)	300	4.00

Note: Center point (+1, −0.4): compression pressure = 300 MPa, disintegrant concentration = 4.00%. 1 e.u. of compression pressure = 100 MPa. 1 e.u. of disintegrant concentration = 2.5%.

is carried out as before, and contour lines are constructed across the new design space.

A variation of the path of steepest ascent method is to construct a line through the center of the design and perpendicular to the contours as before. Then, experiments are performed at intervals along this line, measuring the response at each point. If a maximum or target value is found at a point along the line, that point could serve as the center for a new experimental design.

The new experimental design need not necessarily take the same form as the original. For example, if an inflection point, that is, a maximum or a minimum, were to be found along the path of steepest ascent, then a more complex design, for example, a central composite design, with its center at point A would be more appropriate than the two-level design shown in Figure 8.3.

8.3 OPTIMIZATION BY COMBINING CONTOUR PLOTS

The preceding treatment deals with the two dependent variables separately and hence gives the extreme values of the disintegration time and crushing strength that can be obtained within the given experimental domain. However, these values are not optimal. Experimental conditions that give short disintegration times also give weak tablets, and hence a compromise solution, involving both dependent variables, must be sought.

Figure 8.1 and Figure 8.2 are plotted on identical axes and hence can be super-imposed. This gives a "window" in which is contained all permissible combinations of disintegrant concentration and compression pressure which yield tablets complying with the imposed constraints of dependent and independent variables. This process is facilitated by drawing these graphs on transparent sheets and superimposing them.

Calculations of the type above can be used to ascertain the maximum (or minimum) values of a dependent variable, given certain constraints. A more usual approach is to specify acceptable ranges of values for the dependent variables and then attempt to ascertain the values of the independent variables needed to meet that specification. Suppose that it is required to produce tablets, the disintegration time of which does not exceed 600 sec and which should have a crushing strength in excess of 6 kg, the process being subject to the same constraints as before. The solution space is represented by the hatched portion of Figure 8.4. Any combination of compression pressure and disintegrant concentration lying in this area should give tablets with the specified properties. Thus, the combinations shown in Table 8.3, all obtained without extrapolation, should suffice. For a formulation containing −1.0 e.u. of disintegrant (2.5%), compression should take place within the limited range of −1.05 to +0.05 e.u. (95 to 205 MPa). At higher disintegrant concentrations, for example, +0.50 e.u. (6.2%), a wider range of compression pressures (−0.50 to +1.25 e.u., 150 to 325 MPa) will yield tablets meeting the specifications.

Which of these combinations is chosen will depend on other factors. For example, it would be best to avoid combinations which are at or near the constraints, or which give tablets whose properties are near the specification limits. Alternatively, a cost criterion may be appropriate. This would not be applicable in the present case but could apply when both independent variables are concentrations of two of the ingredients. In such circumstances, the cheapest combination would be chosen.

It is often more important to know whether slight and perhaps inadvertent changes in the values of the independent variables can change the responses and perhaps give rise to an out-of-specification product. This is a measure of the "robustness" of the formulation or process.

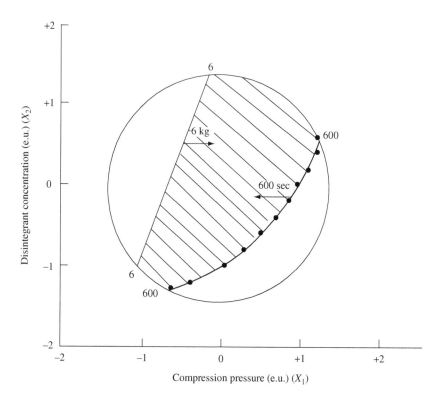

FIGURE 8.4 Composite contour plot to obtain combinations of compression pressure and disintegrant concentration that will give tablets of minimum strength 6 kg and maximum disintegration time of 600 sec.

Of equal importance is the use of composite contour plots such as Figure 8.4 to ascertain whether specifications are feasible. For example, a specification that tablets should disintegrate in less than 200 sec and have a crushing strength of not less than 8 kg cannot be achieved, because there is no combination of disintegrant concentration and pressure that will yield such tablets while remaining within experimental constraints (Figure 8.5). Thus, the specifications of one or both responses must be relaxed.

None of the combinations shown in Table 8.3 is the optimum, and in practice, the precise position of the optimum is often of little importance. However, sometimes, a precise optimum is required. One way to obtain this is to progressively reduce the solution space of Figure 8.4 by moving one or both of the two boundaries, that is, tablet crushing strength and disintegration time. Which of these is moved depends on the perceived relative importance of the two responses. Thus, a target crushing strength of greater than 7 kg could be investigated, keeping the target disintegration time at less than 600 sec.

A similar approach can be used to optimize two responses derived from three factors. Only two factors can be shown on a contour plot. Therefore, the factor that

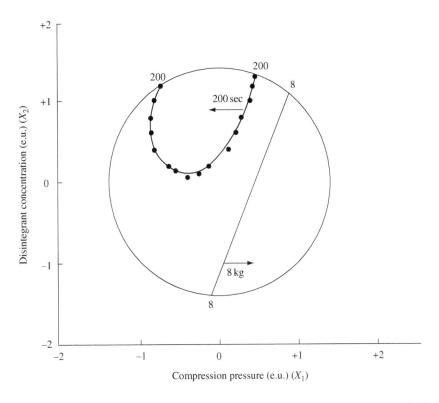

FIGURE 8.5 Composite contour plot showing the impossibility of obtaining tablets of minimum strength 8 kg and maximum disintegration time of 200 sec.

appears to have the smallest effect on the responses is identified and the response surface plotted at specified values of that factor.

8.4 LOCATION OF THE OPTIMUM OF MULTIPLE RESPONSES BY THE DESIRABILITY FUNCTION

The superimposition of contour plots, as described above, identifies an area in which the values of each response are acceptable, for example, a tablet crushing strength greater than 6 kg and a tablet disintegration time not exceeding 600 sec. Every point within the hatched area of Figure 8.4 produces tablets that meet these criteria and, hence, are all equally acceptable.

If a more precise location of the optimum is needed, some form of weighting of the responses is required. One method of achieving this is to use the desirability function introduced by Derringer and Suich.[2] To each response (Y_1, Y_2, etc.) is attached a target value (T_1, T_2, etc.) and a value (or values) which are unacceptable (U_1, U_2, etc.). If a particular response is to be maximized, then if the target value is equalled or exceeded, that response is assigned a partial desirability function (d) of 1. An unacceptable response is assigned the value zero, and if the response lies

TABLE 8.3
Combinations of Disintegrant Concentration and Compression Pressure that will Give Tablets with a Crushing Strength Greater than 6 kg and a Disintegration Time of Less than 600 sec

Disintegrant Concentration		Compression Pressure Range	
e.u.	%	e.u.	MPa
−1.25	1.9	−0.75 to −0.45	125 to 155
−1.00	2.5	−1.05 to +0.05	95 to 205
−0.75	3.1	−0.95 to +0.30	105 to 230
−0.50	3.7	−0.85 to +0.60	115 to 260
−0.25	4.3	−0.80 to +0.85	120 to 285
0.00	5.0	−0.70 to +0.95	130 to 295
+0.25	5.6	−0.60 to +1.15	140 to 315
+0.50	6.2	−0.50 to +1.25	150 to 325
+0.75	6.8	−0.40 to +1.10	160 to 310
+1.00	7.5	−0.30 to +0.90	170 to 290
+1.25	8.1	−0.20 to +0.55	180 to 255

between unacceptability and the target, the partial desirability function lies between 0 and 1 and is calculated from (8.11).

$$d_n = \frac{(Y_n - U_n)}{(T_n - U_n)} \qquad (8.11)$$

If a minimum response is sought, then the partial desirability function of 1 is applied to all responses equal to or less than the target, and values between the target and the unacceptable responses are calculated from (8.12)

$$d_n = \frac{(U_n - Y_n)}{(U_n - T_n)} \qquad (8.12)$$

A situation may also arise whereby there are unacceptable responses both below and above the target value. The partial desirability function may fall below 1 for any deviation from the target, or there may be a range over which all responses are "on target," but outside this range the partial desirability function eventually declines to zero. Figure 8.6(a)–(d) shows these situations in diagrammatic form.

The partial desirability functions for each individual response (d_1, d_2, etc.) can now be combined to give an overall desirability function (D) using (8.13)

$$D = (d_1 \times d_2 \times \cdots \times d_n)^{1/n} \qquad (8.13)$$

D is the geometric mean of all the partial functions and has a range of values from 0 to 1. The geometric mean is used here because if any of the partial desirability functions has a value of zero (i.e., an individual response is unacceptable), then D

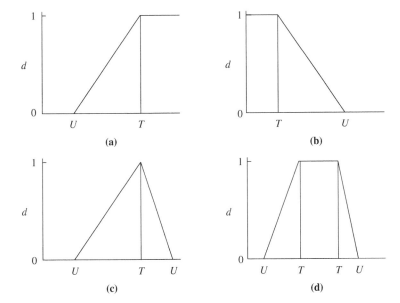

FIGURE 8.6 The relationships between the partial desirability function (d), the target response (T), and the unacceptable response (U): (a) response to be maximized, (b) response to be minimized, (c) single point acceptability, and (d) response acceptable over a range of values.

will also have a value of zero. Use of the arithmetic mean would permit the calculation of a positive value of D, even though one or more of the individual responses were unacceptable.

The desirability technique can be applied to the tablet crushing strength and disintegration time data described earlier. The crushing strength (response Y_1) is to be maximized, with a target level (T_1) of 8 kg or more and an unacceptable level (U_1) of 6 kg or less. The disintegration time (response Y_2) is to be minimized, with a target level (T_2) of 300 sec or less and an unacceptable level (U_2) of 600 sec or more.

The design space is divided into a series of squares, each square representing a convenient increment in both X_1 and X_2, for example, 0.25 e.u. The responses are now calculated from (8.4) and (8.7), respectively. For example, when $X_1 = -0.5$ and $X_2 = -0.5$, the crushing strength given by (8.4) is 6.65 kg and the disintegration time given by (8.7) is 332 sec. Both of these responses are acceptable, but they are considerably short of the target values.

The partial desirability function (d_1) of the crushing strength, given by (8.11), is

$$d_1 = \frac{6.65 - 6}{8 - 6} = 0.323$$

and that for the disintegration time, given by (8.12), is

$$d_2 = \frac{600 - 332}{600 - 300} = 0.892$$

TABLE 8.4

Combined Desirability Function of Tablet Crushing Strength and Disintegration Time Data

		X_1 (e.u.)										
		−1.25	−1.00	−0.75	−0.50	−0.25	0	+0.25	+0.50	+0.75	+1.00	+1.25
	−1.25	0	0.098	0.162	0.071	0	0	0	0	0	0	0
	−1.00	0	0.148	0.323	0.404	0.390	0.158	0	0	0	0	0
	−0.75	0	0	0.356	0.504	0.571	0.544	0.343	0	0	0	0
	−0.50	0	0	0.322	0.536	0.656	0.698	0.640	0.366	0	0	0
	−0.25	0	0	0.203	0.498	0.674	0.778	0.785	0.657	0.279	0	0
X_2 (e.u)	0	0	0	0	0.415	0.615	0.765	0.865	0.829	0.597	0	0
	+0.25	0	0	0	0.312	0.551	0.714	0.846	0.912	0.770	0.456	0
	+0.50	0	0	0	0.150	0.478	0.660	0.801	0.921	0.887	0.651	0
	+0.75	0	0	0	0	0.392	0.600	0.752	0.879	0.959	0.774	0.428
	+1.00	0	0	0	0	0.281	0.534	0.701	0.835	0.951	0.855	0.587
	+1.25	0	0	0	0	0.061	0.488	0.645	0.789	0.910	0.907	0.671

Therefore, the overall desirability function, given by (8.13), is

$$(0.323 \times 0.892)^{1/2} = 0.536$$

Table 8.4 shows the desirability functions for the tablet crushing strength and disintegration time data, using increments of 0.25 e.u. for each factor. The highest value of the combined desirability function is 0.959. This is located at a compression pressure of +0.75 e.u. (275 MPa) and a disintegrant concentration of +0.75 e.u. (6.8%). If a more precise location of the optimal point is needed, then that part of the design space in which the optimum is probably situated can be divided into squares using smaller increments of the factors. Thus, if each square represents an increment of 0.05 e.u., then the optimum is found to be located at a compression pressure of +0.75 e.u. (275 MPa) and a disintegrant concentration of +0.85 e.u. (7.1%). The combined desirability function here is 0.969.

8.5 OPTIMIZATION USING PARETO-OPTIMALITY

Pareto-optimality (named after Vilfredo Pareto, an Italian economist and sociologist) is another technique that uses model equations to locate optimal values.

Using the experimental data given in Table 8.1, regression equations (8.4) and (8.7) are derived as before. Then, using these equations, the values of the dependent variables are calculated for specific values of the independent variables.

Thus, for example, taking the (+1, +1) point of the design, where the compression pressure $(X_1) = +1$ e.u. (300 MPa) and the disintegrant concentration $(X_2) = +1$ e.u. (7.5%), substitution into (8.4) gives a value of crushing strength of 8.22 kg. Similarly, substitution into (8.7) gives a disintegration time of 380 sec. In this way, a pair of results for crushing strength and disintegration time is obtained. This is repeated for other values of compression pressure and disintegrant concentration. In this example, it would be convenient to use intervals of 0.25 e.u. (25 MPa) for compression pressure and 0.25 e.u. (about 0.6%) for disintegrant concentration, which would give 101 pairs of results in the area bounded by the original circular design space. One of each pair of results is then plotted against the other, as shown in Figure 8.7.

Consider any point P on Figure 8.7. The graph can be divided through P into four quadrants, designated I to IV. The objectives of this experiment were to make tablets which have as short a disintegration time as possible and as high a crushing strength as possible. Consider now what these four quadrants signify in relation to P and the experimental objectives.

1. Quadrant I: In relation to P, any points lying in this area have a shorter disintegration time but a lower crushing strength. They would therefore be inferior to P.
2. Quadrant II: In relation to P, any points lying in this area have a longer disintegration time and a lower crushing strength. They would therefore be inferior to P.

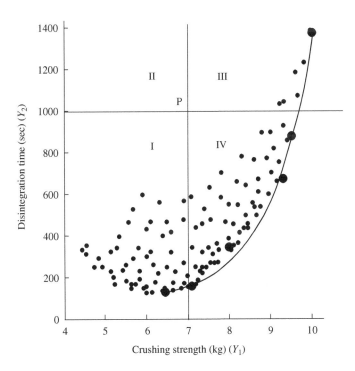

FIGURE 8.7 Pareto-optimal plot of the relationship between tablet disintegration time and crushing strength using data from Table 8.1 and results derived from (8.4) and (8.7). Pareto-optimal points are shown with a larger symbol.

3. Quadrant III: In relation to P, any points lying in this area have a longer disintegration time and a higher crushing strength. They would therefore be inferior to P.
4. Quadrant IV: In relation to P, any points lying in this area have a shorter disintegration time and a higher crushing strength. Therefore, for the purposes of this experiment, points lying in Quadrant IV would be superior to P.

The process can now be repeated, using every other point as a substitute for P until all inferior points have been eliminated. Only the superior or Pareto-optimal points remain. A point is Pareto-optimal if there exists no other point in that space which yields an improvement in one criterion or response without causing a deterioration in another. If a superior point is designated as point P, there are then no points in Quadrant IV.

The Pareto-optimal points are shown with a different symbol in Figure 8.7. Their values and those of the independent variables that give them are shown in Table 8.5.

An advantage of the Pareto-optimal method is that all values of responses within the space covered by the experiment are used. It is not necessary to select predetermined values of the responses, for example, a tablet crushing strength greater than 6 kg. This technique gives more than one "optimum," and the experimenter must then choose which of these is the most acceptable. For example, some points may give tablets with an unacceptably low crushing strength, or where disintegration times exceed compendial standards. These will therefore be rejected.

The Pareto-optimal method can be applied to regression equations derived from the responses of any experimental design. Pareto-optimality has been comprehensively reviewed by Cohon[3] and has been used in a study of tablet formulation by de Boer et al.[4]

FURTHER READING

Response surface methodology and model-dependent optimization have been applied to a wide range of pharmaceutical situations. Reference is made to many review articles, followed by a selected bibliography. Much of the bibliography of Chapter 7 also deals with model-dependent optimization.

TABLE 8.5
Pareto-Optimal Points Obtained from Figure 8.7

Crushing Strength (kg) (Y_1)	Disintegration Time (sec) (Y_2)	Compression Pressure (MPa) (X_1)	Disintegrant Concentration (%) (X_2)
6.5	151	175	6.3
7.2	197	250	8.1
8.4	420	275	5.6
9.1	676	325	5.6
9.5	965	325	3.7
10.0	1368	325	1.9

Reviews

Lewis, G. A., Optimisation methods, in *Encyclopaedia of Pharmaceutical Technology*, Swarbrick, J. and Boylan, J. C., Eds., Dekker, New York, vol. 2, 2002, pp. 1922–1937.

Gonzalez, A. G., Optimization of pharmaceutical formulations based on response-surface experimental designs, *Int. J. Pharm.*, 97, 149, 1993.

Schwartz, J. B., Optimization techniques in product formulation, *J. Soc. Cosmet. Chem.*, 32, 287, 1981.

Sucker, H., Use of optimization techniques in pharmaceutical development, *Drug Dev. Ind. Pharm.*, 15, 1021, 1989.

Appel, L. E., Clair, J. H., and Zentner, G. M., Formulation and optimization of a modified microporous cellulose acetate latex coating for osmotic pumps, *Pharm. Res.*, 9, 1664, 1992.

Bodea, A. and Leucuta, S. E., Optimisation of propranolol hydrochloride sustained release pellets using a factorial design, *Int. J. Pharm.*, 154, 49, 1997.

Bonelli, D. et al., Chemometric modelling of dissolution rates of griseofulvin from solid dispersions with polymers, *Drug Dev. Ind. Pharm.*, 15, 1375, 1989.

Carlotti, M. E. et al., Optimization of emulsions, *Int. J. Cosmet. Sci.*, 13, 209, 1991.

Ceschel, G. C., Maffei, P., and Badiello, R., Optimisation of a tablet containing chlorthalidone, *Drug Dev. Ind. Pharm.*, 25, 1167, 1999.

Costa, F. O. et al., Comparison of dissolution profiles of ibuprofen tablets, *J. Control. Rel.*, 89, 199, 2003.

Dawoodbhai, S., Suryanarayan, E. R., and Woodruff, C. W., Optimization of tablet formulations containing talc, *Drug Dev. Ind. Pharm.*, 17, 1343, 1991.

Diemunsch, A. M. et al., Tablet formulation: Genichi Taguchi's approach, *Drug Dev. Ind. Pharm.*, 19, 1461, 1993.

Hauer, B., Remmele, T., and Sucker, H., Rational development and optimization of capsule formulations with an instrumented dosator capsule filling machine. Part 2: Fundamentals of the optimization strategy, *Pharm. Ind.*, 55, 780, 1993.

Johnson, A. D., Anderson, V. L., and Peck, G. E., Statistical approach for the development of an oral controlled-release tablet, *Pharm. Res.*, 7, 1092, 1990.

Lemaitre-Aghazarian, V. et al., Texture optimization of water in oil emulsions, *Pharm. Dev. Technol.*, 9, 125, 2004.

Marengo, E. et al., Scale-up and optimization of an evaporative drying process applied to aqueous dispersions of solid lipid nanoparticles, *Pharm. Dev. Technol.*, 8, 299, 2003.

Martinez, S. C. et al., Aciclovir poly(D, L-lactide–co-glyceride) microspheres for intravitreal administration using a factorial design study, *Int. J. Pharm.*, 273, 45, 2003.

McGurk, J. G., Lendrem, D. W., and Potter, C. J., Use of statistical experimental design in laboratory scale formulation, optimization and progression to plant scale, *Drug Dev. Ind. Pharm.*, 17, 2341, 1991.

Pena-Romero, A. et al., Statistical optimization of a sustained release form of sodium diclofenac on inert matrices. Part 2. Statistical optimization, *Pharmaceutica Acta Helvetiae*, 63, 333, 1988.

Senderak, E., Bonsignore, H., and Mungan, D., Response surface methodology as an approach to optimization of an oral solution, *Drug Dev. Ind. Pharm.*, 19, 405, 1993.

Shirakura, O. et al., Particle size design using computer optimization techniques, *Drug Dev. Ind. Pharm.*, 17, 471, 1991.

Takayama, K. et al., Formulation design of indomethacin gel ointment containing D-limonene using computer optimization methodology, *Int. J. Pharm.*, 61, 225, 1990.

Vojnovic, D. et al., Simultaneous optimization of several response variables in a granulation process, *Drug Dev. Ind. Pharm.*, 19, 1479, 1993.

Wehrle, P. et al., Response surface methodology: interesting statistical tool for process optimization and validation: example of wet granulation in a high-shear mixer, *Drug Dev. Ind. Pharm.*, 19, 1637, 1993.

REFERENCES

1. Myers, R. H. and Montgomery, D. C., *Response Surface Methodology: Process and Product Optimisation using Designed Experiments*, 2nd ed., Wiley, New York, 2002.
2. Derringer, G. and Suich, R., Simultaneous optimisation of several response variables, *J. Qual. Tech.*, 12, 214, 1980.
3. Cohon, J. L., *Multiobjective Programming and Planning*, Academic Press, New York, 1978.
4. De Boer, J. H., Smilde, A. K., and Doornbos, D. A., Introduction of multi-criteria decision making in optimization procedures for pharmaceutical formulations, *Acta Pharm. Technol.*, 34, 140, 1988.

9 Sequential Methods and Model-Independent Optimization

9.1 INTRODUCTION

In most experiments, the experiment is designed, then carried out, and only after completion are the accumulated data evaluated. It is thus possible to predict the amount of data that will be amassed even before experimentation has been started. Such experiments are therefore called fixed sample tests. A disadvantage of such a design is that a significant result could remain unnoticed until data collection is complete.

In sequential methods, results are continually examined as they become available. Most applications of sequential analysis have been in the medical field, when it is important to know as soon as possible whether a significant result has been obtained. The trial can then be stopped and all patients given the successful treatment. Indeed, it would be unethical to do otherwise. However, there is no reason to limit sequential methods to medical trials, and they form a useful method of goal seeking and optimization.

9.2 SEQUENTIAL ANALYSIS

The requisites for sequential analysis are two treatments, termed A and B, and criteria for success or failure.

9.2.1 WALD DIAGRAMS

This is the most straightforward technique in sequential analysis. It is described briefly here, but for full mathematical details, together with other sequential analytical techniques, the reader is referred to Whitehead.[1] The experiments are carried out in pairs, one of each pair receiving treatment A and the other treatment B. There is thus a series of individual small trials, each involving a paired comparison.

For a given pair of experiments, there are four possible outcomes:

1. Both treatments are successful
2. Both treatments fail
3. Treatment A succeeds and treatment B fails
4. Treatment B succeeds and treatment A fails.

1 and 2 are called tied pairs, and though they may provide valuable information, they are not used in sequential analysis.

The technique is illustrated using data shown in Table 9.1.

These data were originally obtained by Brown et al.,[2] who investigated the effects of two doses of antitoxin in clinical tetanus. A high dose is designated treatment A and a low dose treatment B. The criteria for success or failure were the survival or the death of the patient. The necessity of obtaining a conclusive result as soon as possible is apparent!

Although Table 9.1 shows 25 pairs of results, each pair is examined as it becomes available. The results are plotted on a Wald diagram[3] (Figure 9.1). The horizontal axis represents the number of untied pairs $(n_A + n_B)$ and the vertical axis the results of paired comparisons. Zero is halfway up the vertical axis. A positive scale represents pairs in which treatment A succeeds, and a corresponding negative scale represents failure for treatment A, that is, success for treatment B.

The plot starts at the origin. Success for treatment A is represented by a line with a positive slope (/), drawn across the first square. Success for treatment B is represented by a line drawn in the opposite direction (\). The procedure is carried out for each successive pair, the new line starting where the previous one ended. With the data given in Table 9.1, the results give a zigzag line with an overall positive slope, indicating that treatment A is superior to treatment B. Had the line gone in the other direction, the superiority of treatment B would be indicated. If an

TABLE 9.1
Paired Results for Treatments A and B

Pair	Treatment	
	A	B
1	Success	Failure
2	Success	Failure
3	Failure	Success
4	Success	Failure
5	Success	Failure
6	Success	Failure
7	Success	Failure
8	Failure	Success
9	Success	Failure
10	Success	Failure
11	Failure	Failure
12	Success	Failure
13	Success	Failure
14	Success	Failure
15	Success	Success
16	Failure	Failure
17	Success	Failure
18	Failure	Success
19	Success	Failure
20	Success	Failure
21	Success	Failure
22	Failure	Failure
23	Failure	Success
24	Success	Failure
25	Failure	Success

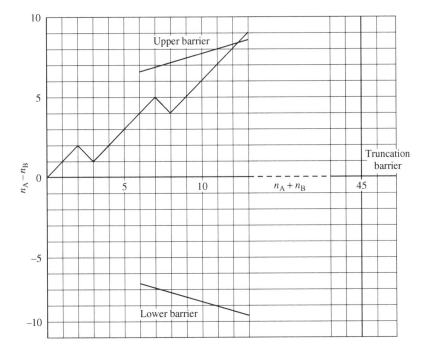

FIGURE 9.1 Wald diagram of the data presented in Table 9.1.

approximately horizontal line had been followed, the inference would be that neither treatment was superior.

The solid lines above and below the plot and on the right constitute barriers or boundaries. If the plot of the original results crosses the upper boundary, this indicates that treatment A is significantly better, and if the plot crosses the lower boundary, then treatment B is significantly better. If the end or truncation barrier is crossed, then the experiment is inconclusive.

The positions of the barriers are thus crucial. The upper and lower barriers are constructed using the principles of the paired sign test described in Chapter 3. If there is no significant difference between items of paired data, then the difference between the two items in a pair is as likely to be positive as it is to be negative. Thus, the vertical scale of a Wald diagram represents the number of untied pairs favoring treatment A (n_A) minus the number of untied pairs favoring treatment B (n_B), that is, $n_A - n_B$.

Figure 9.2 represents the upper part of Figure 9.1 in more detail.

The numbers given in Table 9.2 are derived from the paired sign test discussed in Chapter 3. The "excess positives" are the number of positive untied pairs to give significance at $P = 0.05$. Thus, for a sample size ($n_A + n_B$) of 6, all results would have to be positive, whereas for a sample size of 12, the number of excess positives for significance would be 8.

The upper boundary of Figure 9.2 is obtained by plotting $n_A + n_B$ from Table 9.2 as the abscissa against the corresponding value of $n_A - n_B$. Because of their shape, these barriers are often called 'Christmas tree' boundaries. It is common practice to

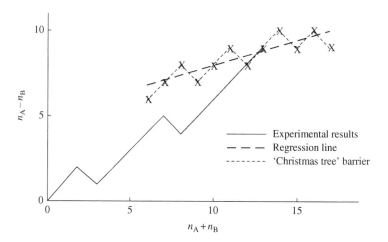

FIGURE 9.2 Derivation of the barriers of a Wald diagram, using data presented in Table 9.1.

plot the best straight line through these points and use the regression line as the boundary. Linear regression analysis of excess positives ($n_A - n_B$) against the total number of untied pairs ($n_A + n_B$) gives (9.1)

$$(n_A - n_B) = 0.279(n_A + n_B) + 5.11 \qquad (9.1)$$

The coefficient of determination (r^2) of this line is 0.807.

TABLE 9.2
Excess Positives for Sequential Analysis ($P = 0.05$)

Sample Size ($n_A + n_B$)	Number of Positives (n_A)	Number of Negatives (n_B)	Excess Positives ($n_A - n_B$)
6	6	0	6
7	7	0	7
8	8	0	8
9	8	1	7
10	9	1	8
11	10	1	9
12	10	2	8
13	11	2	9
14	12	2	10
15	12	3	9
16	13	3	10
17	13	4	9
18	14	4	10
19	15	4	11
20	15	5	10

Note: n_A represents the number of untied pairs favoring A and n_B the number favoring B.

The barrier line for the lower half of Figure 9.1 is derived in the same way, but with a negative slope and intercept, as shown in (9.2)

$$(n_A - n_B) = -0.279(n_A + n_B) - 5.11 \qquad (9.2)$$

Thus, when the line formed by plotting the untied pairs of results crosses either the upper or lower boundary lines, a significant result is obtained at $P = 0.05$. If this level of significance is satisfactory, the trial can then be terminated.

The possibility that neither the top nor the bottom boundary is crossed must now be considered. This may happen for two reasons. Either there is no difference between the treatments, in which case neither boundary will ever be crossed. Or there is a difference, but it is small, and therefore significance will only be established after many experiments. For this reason, the boundary on the right side of the Wald diagram is constructed by a process called truncation. The first stage of this process is to calculate the "longest path" from (9.3)

$$\text{longest path} = k(\text{number of paired experiments}) \qquad (9.3)$$

This total includes all tied as well as untied pairs. In turn, k is given by (9.4)

$$k = \frac{1}{p_B(1 - p_A) + p_A(1 - p_B)} \qquad (9.4)$$

where
p_B = the proportion of successes expected from treatment B
p_A = the proportion of successes of treatment A, which would be considered a worthwhile improvement over treatment B.

A judgment is taken *before the experiment is started* over what an acceptable response would be. For example, if treatment B would be expected to succeed on 40% of occasions, $p_B = 0.4$. If a success rate of at least 80% for treatment would make treatment A worthwhile, then $p_A = 0.8$. Substitution into (9.4) gives a value of k of 1.785. Then, substitution into (9.3) gives a longest path of 45. A vertical barrier is then drawn on Figure 9.1 at this value and the experiment stopped if the upper and lower boundaries are not reached by this point. Other truncation techniques are described by Whitehead.[1]

All truncation methods are arbitrary and are therefore not totally satisfactory. Indeed, it can be argued that truncation is not necessary at all. If the number of pairs (25 in this case) is exhausted without crossing an upper or lower barrier, then the trial becomes a fixed sample test, and decisions can be made on the basis of traditional significance test methods.

9.3 MODEL-INDEPENDENT OPTIMIZATION

9.3.1 OPTIMIZATION BY SIMPLEX SEARCH

In the model-dependent methods described in Chapter 8, a series of experiments is designed, the experiments carried out, and only when all experiments have been completed is a model devised. On the other hand, the simplex search method, developed by Spendley et al.,[4] is an optimization procedure that adopts a more empirical sequential approach. The results of previous experiments are used in a

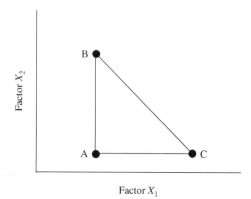

FIGURE 9.3 The first stage in optimization by simplex search.

mathematically rigorous manner to define the experimental conditions of subsequent experiments in an attempt to find the optimal response. The optimum is approached by moving away from the undesirable values of the response.

The name simplex derives from the shape of the geometric figure that moves across the response surface. It is defined by the number of vertices equal to one more than the number of variables in the space. Thus, a simplex of two variables is a triangle.

The basis of the method is most readily grasped in the case where there are two independent variables: X_1 and X_2. The simplex is constructed by selecting three combinations of these two variables. These combinations are designated A, B, and C. The three experiments represented in Figure 9.3 are carried out, and the response is measured in each case. These responses are designated R_A, R_B, and R_C, respectively.

It must be decided at the outset whether the desired goal is a maximum (e.g., tablet crushing strength which should be as high as possible) or a minimum (such as tablet disintegration time which should be as short as possible). In the next few paragraphs and in Table 9.3, the terms "better" or "worse" are used rather than "greater" or "less." "Better" implies progress toward the goal, be that a maximum or a minimum.

Let us assume that the response at A is worse than those at B and C. The values of the independent variables for the next experiment (D) are therefore chosen by moving away from point A. This is achieved by reflecting the triangle ABC about the BC axis. Hence, AP=DP. The experiment at point D is performed and the response R_D compared with the responses at points A, B, and C (Figure 9.4).

TABLE 9.3

Procedure to Determine the Course of Action in a Simplex Search after Responses are Obtained at Points A, B, C, and D

Relative Value of Response	Course of Action
R_D better than R_A, R_B, and R_C	Expand further along line APD to point E (Figure 9.5a)
R_D better than R_A and R_B, but worse than R_C	Reflect triangle BCD about CD axis to point F (Figure 9.5b)
R_D better than R_A, but worse than R_B and R_C	Contract along line PD to point G (Figure 9.5c)
R_D worse than R_A, R_B, and R_C	Contract along line AP to point H (Figure 9.5d)

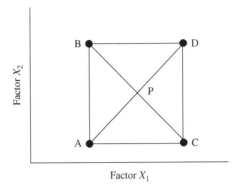

FIGURE 9.4 The second stage in optimization by simplex search.

The next move depends on the relative values of the four responses:

1. If R_D is better than R_A, R_B, or R_C, then it is worthwhile proceeding further in the AD direction. The next point E is located along this line such that PD = DE (Figure 9.5a). This procedure is termed expansion.
2. If R_D is better than R_A and R_B but worse than R_C, then vertex D is retained, and the next point F is located by moving away from B, reflecting triangle BCD about axis CD (Figure 9.5b).
3. If R_D is worse than R_B and R_C, but better than R_A, the next experiment (G) is located along the AD axis at (P+0.5AP) (Figure 9.5c).
4. Lastly, if R_D is worse than R_A, R_B, or R_C, then point H is located along the same axis at (P−0.5AP) (Figure 9.5d).

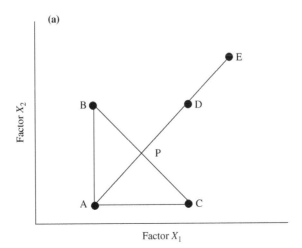

FIGURE 9.5 [(a)–(d)] Alternative courses of action in subsequent stages in optimization by simplex search.

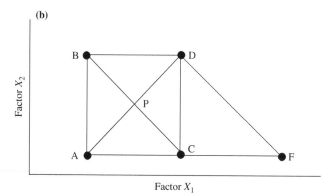

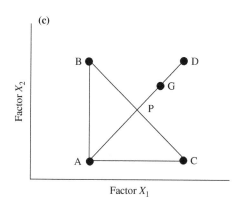

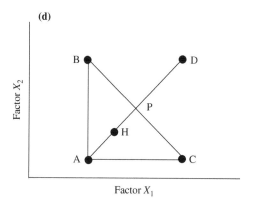

FIGURE 9.5 [(a)–(d)] continued

The last two procedures are known as contractions.

The overall position is summarized in Table 9.3.

The procedure is then repeated, comparing the result of the latest experiment (i.e., E, F, G, or H) with those that have gone before and positioning further experiments according to the strategies laid down in Table 9.3. It is likely that the position of some experiments will violate the boundaries or constraints of the design space and hence cannot be used.

An example of how the simplex approach can be applied is provided by the work of Gould and Goodman.[5] They used the technique to determine the blend of ethanol, propylene glycol, and water in which caffeine showed the greatest solubility. The objective of this series of experiments is thus to maximize solubility.

The three initial combinations of ethanol and propylene glycol are designated Vertices 1, 2, and 3, respectively, in Table 9.4. Of these, Vertex 3 gives the lowest solubility, and hence Vertex 4 is located by reflection along the 1–2 axis. Solubility at Vertex 4 is higher than that at Vertices 2 and 3, and hence further expansion along the 1–4 axis to Vertex 5 is probably worthwhile. However, this combination gives the lowest solubility of all, and hence further progression along this line is pointless. Consideration, thus, returns to Vertices 1, 2, and 4. Of these, Vertex 1 is the lowest, and hence reflection from that point about the 2–4 axis gives Vertex 6.

The triangle formed by Vertices 2, 4, and 6 is now considered. Vertex 2 is the lowest of these three points, so reflection now occurs about the 4–6 axis to give Vertex 7. Solubility at Vertex 7 is lower than that at both Vertices 4 and 6 but higher than at Vertex 2, so contraction now occurs to give Vertex 8, and then the experiment at Vertex 9 is carried out. The last two points give virtually the same result, indicating that a maximum is nearby. The precise point of the maximum could be found by further experiments if this is considered worthwhile. The sequence of experiments is shown in Figure 9.6.

TABLE 9.4
Vertices, Solvent Blends, and the Solubility of Caffeine in Those Solvent Blends

Vertex	Ethanol (%v/v)	Propylene Glycol (%v/v)	Solubility (mg·ml⁻¹)	Vertices Retained	Vertex Rejected	Process
1	0	40	24.0	–	–	–
2	20	0	26.2	–	–	–
3	0	0	17.2	–	–	–
4	20	40	44.9	1, 2	3	Reflection
5	30	60	17.5	1, 2	3	Expansion
6	40	0	52.4	2, 4	1	Reflection
7	40	40	36.7	4, 6	2	Reflection
8	35	30	52.9	4, 6	7	Contraction
9	29	28	53.0	4, 6	8	Contraction

From Gould, P. L. and Goodman, M., *J. Pharm. Pharmacol.*, 35, 3P, 1983. With permission.

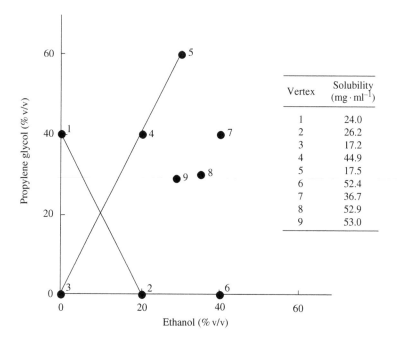

FIGURE 9.6 Optimization by simplex search. The numbers on the graph refer to the vertices shown in Table 9.4. (From Gould, P. L. and Goodman, M., *J. Pharm. Pharmacol.*, 35, 3P, 1983. With permission.)

As only one response, solubility, is considered in this series of experiments, the outcome is the maximum solubility, not an optimum value. At least two responses are required for optimization. Also, it must be noted that though the authors consider changes in the concentration of only two of the liquids, a third (water) is present, the concentration of which is unavoidably altered if the sum of the other two ingredients is changed. Optimization of mixtures, in which the sum of the proportions of all the ingredients totals unity, is dealt with in more detail in Chapter 10.

A true example of optimization by the simplex approach is given by the work of Shek et al.,[6] who investigated factors involved in capsule formulation. Four independent variables were chosen, namely, the concentrations of drug, disintegrant, and lubricant, and the total capsule weight. As there are four independent variables, the simplex for this design is a pentagon. Shek et al. decided that there were three responses of interest, namely, rate of packing down or consolidation of the powder (R_1), percentage of drug dissolved at 30 min (R_2), and percentage of drug dissolved at 8 min (R_3). The first of these was determined using a mechanical tapping device, and the units are the number of taps required to achieve the final volume of the powder. Because rapid consolidation of the powder was considered to be desirable, the number of taps should be as low as possible. Dissolution rate should be as rapid as possible, and thus the maximum percentage dissolved in a specific time is required. Thus, an optimum solution is sought.

The units in which both independent and dependent variables are expressed must be considered. In this example, three of the four independent variables have units of percentage concentration, but the regions of interest that they cover are very different. For example, Shek et al. used disintegrant concentrations between 0% and 50%, whereas the lubricant concentration ranged between 0% and 2.2%. The fourth independent variable has a completely different unit, namely, weight, and covers a range of several hundred milligrams. All four variables must therefore be put on the same unitary basis. This is achieved by a process of "normalization." The process is as follows: upper and lower limits of each independent variable are selected. These are designated H and L and are the extreme values of a particular variable, which are likely to be of interest. Their selection is based on experience or on limits imposed by the variables themselves. For example, Shek et al. selected limits of capsule weight to be 100 mg and 400 mg, and these values presumably were derived from the sizes of available capsule shells or filling equipment. This is the equivalent of setting constraints in a model-dependent optimization procedure.

Values of the independent variables are then normalized by (9.5), where N is the normalized value and X is the uncorrected value of that variable

$$N = \frac{[(X-L)\times 100\%]}{(H-L)} \tag{9.5}$$

Thus, a capsule weight of 200 mg, when normalized, would become

$$\frac{[(200-100)\times 100\%]}{(400-100)} = 33.3\%$$

Normalization of the independent variables is not necessary, if all are expressed in the same units and cover the same range. For example, in Gould and Goodman's work discussed earlier, all had units of %v/v and covered the range 0–100%.

As described so far, the simplex search method can be used to independently maximize or minimize each of the three dependent variables or responses. However, to determine the optimum response, two further procedures are needed. The first is to normalize the responses. If the desired response is tending toward as high a value as possible, as with R_2 and R_3 in this case, then (9.5) is used. Where the desired response tends toward as low a value as possible, as with R_1, then (9.6) is used

$$N = \frac{[(H-X)\times 100\%]}{(H-L)} \tag{9.6}$$

This ensures that all normalized values are positive.

The second procedure is to give each individual response a weighting factor that reflects the relative importance of that response to the overall success of the experiment. Shek et al. decided that the three responses should have the relative importance of 0.5, 0.4, and 0.1, respectively. This means that the rate of packing down was considered to be the most important, followed by the percentage of drug dissolved in 30 min. The percentage of drug dissolved in 8 min was viewed as the least important response. Because rapid packing down is considered desirable, the number of taps should be as low as possible. The other two responses are maxima.

Thus, if R_t is the total response, then (9.7) applies

$$R_t = -0.5R_1 + 0.4R_2 + 0.1R_3 \qquad (9.7)$$

Each of the responses is measured and normalized and the total response calculated according to (9.7).

In general terms, if at any given combination of independent variables the responses are R_1, R_2, $R_3, \ldots,$ R_n, and the weighting factors a_1, a_2, $a_3, \ldots,$ a_n respectively, then the total response, R_t, is given by (9.8)

$$R_t = a_1R_1 + a_2R_2 + a_3R_3 + \cdots + a_nR_n \qquad (9.8)$$

where
$a_1 + a_2 + a_3 + \cdots + a_n = 1.$

If all n responses are judged to be of equal importance, then the weighting factor is $1/n$. These weighting factors should be selected before experimentation starts.

The calculated value of R_t has the units of percent, and the precise optimum combination of responses will therefore have a value of 100%. It is unlikely that this would ever be achieved. Among other considerations, it implies that all the H and L values in the normalization process have been selected with total accuracy. Hence, experimentation can be reduced by specifying a lower but acceptably high value for R_t.

9.4 COMPARISON OF MODEL-INDEPENDENT AND MODEL-DEPENDENT METHODS

The simplex approach can be regarded as a step-by-step process of achieving the optimum. It must be conceded that many steps may be needed before that optimum is reached. For example, Gould and Goodman carried out 9 experiments and Shek et al. 45 before a satisfactory optimum was achieved. However, a willingness to settle for less than the precise optimum greatly reduces the number of experiments. Thus, if Gould and Goodman, rather than searching for the maximum solubility, had looked for a solvent mixture in which caffeine was soluble in excess of 50 mg·ml⁻¹, then this would have been discovered in six experiments. Though the numbers of experiments used by Shek et al. may seem high, a full three-level factorial design, using four factors, would necessitate 81 experiments. However, use of some form of fractional design would reduce this number (see Chapter 6). A further point in favor of a sequential approach is that if all experiments are carried out simultaneously, it may be that the experimental design has been devised with inappropriate values for the independent variables. As an example, consider the tablet formulation exercise described in Chapter 8. If the highest compression pressure yielded tablets whose crushing strength exceeded the range of the measuring apparatus, then the design would have to be repeated using lower compression pressures. Even then, derivation of the model might show that the optimum lay outside the chosen ranges of values for the independent variables. This would necessitate extrapolation or repeating the design over other ranges.

However, if one is willing to accept a response that is less than optimal, then it must also be accepted that there may well be many combinations of experimental conditions that will give this. Thus, for example, Gould and Goodman found a solvent blend in which the solubility of caffeine exceeded $50 \, mg \cdot ml^{-1}$ in their sixth experiment. However, this may not be the "best" blend of ethanol, propylene glycol, and water that gives this solubility. An obvious further consideration is cost. If the three components of the mixture differ in cost, as seems likely in this case, then it is sensible to select the cheapest combination that gives the required effect. It may therefore be useful to combine the model-independent simplex approach with a model-dependent technique. Using data derived in the simplex search, regression analysis is carried out, followed by mapping of the response surface. This combined approach was used by Shek et al., and the reader is referred to their article for further details.

However, the model-independent method has disadvantages. Though fewer experiments may be needed to reach an optimal solution, this number is not known at the outset, and hence difficulties in planning the work may arise.

Of more importance is that the simplex search leads to one optimal solution (or one maximum or minimum, depending on the experiment). Nothing is known about other areas of the response surface, and there may be better solutions to the problem in areas that have not been explored. To persist with the analogy of climbing a mountain, reaching a summit does not guarantee that the highest peak in the whole mountain range has been achieved. Neither is anything known about the stability or robustness of the solution. The peak may be a plateau, so that slight deviation from the optimal conditions will have little effect on the response. On the other hand, with a sharp peak, a slight variation will lead to a major change in response. If contour plots are derived as part of a model-dependent design, then the nature of the peak is apparent.

Another potential weakness of the model-independent method is the need to normalize the values of the independent variables and responses and the consequent selection of upper and lower limits. These are bound to be based, in part at least, on informed guesswork.

The use of weighting factors to achieve an optimal solution can also cause problems, because the choice of inappropriate factors can alter the conclusions drawn from the experiment. This can be appreciated from the following example. The solubility data are those presented in Table 9.4, but the aim is now to produce a solution that is optimal with regard to both solubility and cost. For the purposes of this example, the price of one liter of ethanol is assumed to be £10.00, that of propylene glycol £5.00, and that of water £0.50.

Consider the first three experiments in Table 9.4. There is no need to normalize the independent variables as they all have the same units, but the responses must be normalized. The aim is thus to maximize solubility and minimize cost. In

TABLE 9.5
Solvent Blends and Normalized Solubility and Cost Responses, Adapted from Table 9.4

				Solubility		Cost	
Vertex	Ethanol (%v/v)	Propylene Glycol (%v/v)	Water (v/v)	Actual (mg·ml^{-1})	Normalized (%)	Actual (£·liter^{-1})	Normalized (%)
1	0	40	60	24.0	40.0	2.30	81.1
2	20	0	80	26.2	43.7	2.40	74.7
3	0	0	100	17.2	28.7	0.50	94.7

calculating the total response, solubility (R_1) is normalized according to (9.5) and cost (R_2) is normalized according to (9.6).

The chosen range of solubilities is 0 to 60 mg·ml^{-1}, and the cost range is £0.50 to £10.00 per liter, that is, pure water to pure ethanol. The actual and normalized cost and solubility data is given in Table 9.5.

The total response can be calculated according to (9.9).

$$R_t = a_1 R_1 + a_2 R_2 \qquad (9.9)$$

It is instructive to consider three pairs of weighting factors. With the first pair, both responses are considered to be of equal importance. Therefore, $a_1 = a_2 = 0.5$. In the other two pairs, one response is considered to be four times more important than the other. Thus, $a_1 = 0.8$, $a_2 = 0.2$ and $a_1 = 0.2$, $a_2 = 0.8$. The total responses calculated from these three pairs are shown in Table 9.6.

When both responses have equal weighting, or when cost is deemed to be four times as important as solubility, Vertex 2 gives the lowest total response, and hence the position of the next experiment is located by moving away from Vertex 2. Full reflection is not possible because the boundaries of the design would be violated, and hence the next experiment is located at Vertex 4A (Figure 9.7). The composition

TABLE 9.6
Calculation of the Total Response to the Experiments Shown in Table 9.5

	Total Response		
Experiment	$a_1 = 0.5$ $a_2 = 0.5$	$a_1 = 0.8$ $a_2 = 0.2$	$a_1 = 0.2$ $a_2 = 0.8$
1	60.6	48.2	72.9
2	59.2	49.9	68.5
3	61.7	41.9	81.5
Lowest vertex	2	3	2
Position of next experiment	4A	4B	4A

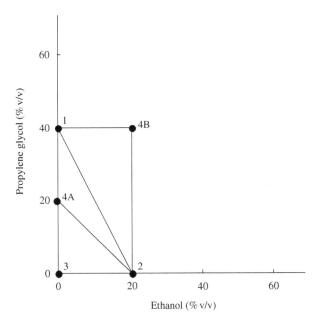

FIGURE 9.7 The effect of choice of weighting factors on the positioning of experiments in a simplex search, using data shown in Table 9.6.

of the solvent blend at this point is 20% propylene glycol and 80% water. On the other hand, if the solubility response is considered to be four times as important as the cost, then Vertex 3 gives the lowest response and hence the next experiment is situated at Vertex 4B, the composition of which is 20% ethanol, 40% propylene glycol, and 40% water. This, in turn, will affect the positioning of all subsequent experiments.

FURTHER READING

The following articles describe the use of model-independent optimization in the design and evaluation of experiments. Reference to two review articles is given, followed by a selected bibliography.

De Boer, J. H., Smilde, A. G., and Doornbos, D. A., Introduction of multi-criteria decision-making in optimization procedures for pharmaceutical formulations, *Acta Pharm. Technol.*, 34, 140, 1988.

Schwartz, J. B., Optimization techniques in product formulation, *J. Soc. Cosmet. Chem.*, 32, 287, 1981.

Armitage, P., *Sequential Medical Trials*, 2nd ed., Blackwell, Oxford, 1975.

Bross, I., Sequential medical plans, *Biometrics*, 8, 188, 1952.

Dols, T. J. and Armbrecht, B. H., Simplex optimization as a step in method development, *J. Assoc. Off. Anal. Chem.*, 59, 1204, 1976.

Gould, P. L., Optimisation methods for the development of dosage forms, *Int. J. Pharm. Technol. Prod. Manuf.*, 5, 19, 1984.

Hamed, E. and Sakr, A., Application of multiple response optimisation techniques to extended release formulations, *J. Control. Rel.*, 73, 329, 2001.

Lewis, A. E., *Biostatistics*, Reinhold, New York, 1966.

Masilungan, F. C., Carabba, C. D., and Bohidar, N. R., Application of simplex and statistical analysis for correction of pitting in aqueous film coated tablets, *Drug Dev. Ind. Pharm.*, 17, 609, 1991.

Masilungan, F. C. and Kraus, K. F., Determination of precompression and compression force levels to minimize tablet friability using simplex, *Drug Dev. Ind. Pharm.*, 15, 1771, 1989.

Thoennes, C. J. and McCurdy, V. E., Evaluation of a rapidly disintegrating moisture resistant lacquer film coating, *Drug Dev. Ind. Pharm.*, 15, 165, 1989.

Wehrle, P., Nobilis, P., and Stamm, A., Study of the lubrication of a soluble tablet. Part 1. Treatment of sodium benzoate for the improvement of its lubricant properties, *STP PHARMA Science*, 4, 202, 1988.

REFERENCES

1. Whitehead, J., *The Design and Analysis of Sequential Clinical Trials*, 2nd ed., Ellis Horwood, Chichester, 1997.
2. Brown, A. et al., Value of a large dose of antitoxin in clinical tetanus, *Lancet*, 2, 227, 1960.
3. Wald, A., *Sequential Analysis*, Wiley, New York, 1948.
4. Spendley, W., Hext, G. R., and Himsworth, F. R., Sequential application of simplex designs in optimisation and evolutionary operations, *Technometrics*, 4, 441, 1962.
5. Gould, P. L. and Goodman, M., Simplex search in the optimisation of the solubility of caffeine in parenteral cosolvent systems, *J. Pharm. Pharmacol.*, 35, 3P, 1983.
6. Shek, E., Ghani, M., and Jones, R. E., Simplex search in optimization of capsule formulation, *J. Pharm. Sci.*, 69, 1135, 1980.

10 Experimental Designs for Mixtures

10.1 INTRODUCTION

Much of the foregoing discussion has considered the effect of altering one or more factors, often environmental factors such as temperature, on the outcome of an experiment. However, pharmaceutical formulations are almost invariably mixtures of active ingredients and excipients. In this case, the experimental response may be a property of that mixture, and the factor is the composition of the mixture.

It is important to distinguish between the proportions of ingredients in a formulation and the actual amounts. For example, in a tablet, the total weight of each unit is not usually fixed. Hence, if the amounts of excipients such as lubricant or disintegrating agent are changed, this will change the final weight of each tablet. There will then be a consequent effect on the proportions of those excipients in the formulation, but it is usually the concentration of excipient in the overall formulation, for example, 0.5% magnesium stearate, which is of interest. On the other hand, the solubility of an active ingredient in a mixture of cosolvents is governed by the composition of the mixture, and aspects of experimental design related to such systems are dealt with in this chapter.

Because the proportions of all components in a mixture must total unity, these proportions are not truly independent. Any alteration in the proportion of one component in the mixture must of necessity change the proportion of at least one other ingredient. Furthermore, each of the proportions of the components must be nonnegative, that is, they must be either zero or a positive number.

Putting the foregoing in general terms, a pharmaceutical formulation can be regarded as a mixture consisting of q components that are the active ingredient(s) and the excipients.

If we designate the proportions of these components $X_1, X_2, \ldots, X_q$, then

$$0 \leq X_i \leq 1$$

where

$X_i =$ any number from 1 to q.

The sum of the proportions of all the components is unity.

Therefore, $X_1 + X_2 + \cdots + X_q = 1$.

The factor space, which is the area representing all possible combinations of the components, can be represented by the interior and the boundaries of a regular figure with q vertices and $q-1$ dimensions. Therefore, all proportions of two ingredients can

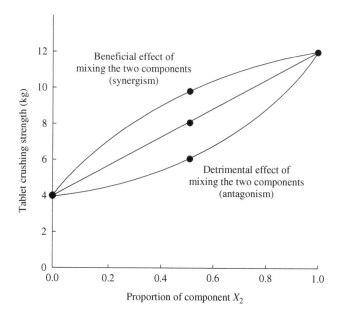

FIGURE 10.1 Tablet crushing strength as a function of the proportions of Components X_1 and X_2.

be represented by a straight line. Consider Figure 10.1, which shows the crushing strength of tablets (the response) as a function of the relative proportions of two solid components X_1 and X_2. The factor space is the line joining the points representing pure X_1 and pure X_2. This is the abscissa of Figure 10.1, and the two "vertices," that is, the ends of the line, are the crushing strengths of pure X_1 and pure X_2.

If the responses to the ingredients are purely additive, then the response line will be the straight line joining the crushing strengths of the two pure diluents. These two points can be regarded as "single component mixtures," and the shape of the line joining them is a useful benchmark to assess whether mixing the two components together has a beneficial or a detrimental effect. An upwardly concave line indicates that the tablets made of a mixture have a lower crushing strength than would have been predicted by simple proportionality, and hence mixing the ingredients has an antagonistic effect. An upwardly convex line indicates synergism, in that the tablets of mixtures are stronger than simple proportionality would predict.

10.2 THREE-COMPONENT SYSTEMS AND TERNARY DIAGRAMS

If there are three components ($q = 3$), then the factor space is represented by a two-dimensional, three-cornered figure, which is an equilateral triangle. From any point within an equilateral triangle, the sum of the distances perpendicular to each side is equal to the height of the triangle. By taking the length of each side as unity, and expressing the amounts of the three components as fractions or proportions of the whole, it is possible to represent the composition of any mixture by a point on Figure 10.2, giving a ternary diagram.

The three components are designated X_1, X_2, and X_3. Each of the three corners of the triangle represents a pure component. Hence, the proportion of that component at that point is 1. Thus, point B represents a formulation consisting entirely of Component X_2, Components X_1 and X_3 being absent. The boundaries of the triangle, being straight lines, represent two-component systems. Thus, the base of the triangle represents all possible mixtures of Component X_1 and Component X_3. Point D, which is halfway along this line, represents a mixture containing equal proportions of X_1 and X_3, Component X_2 being absent. The scale for all three sides must be the same. It is usual, though not essential, for the scales to increase in a clockwise direction rather than anticlockwise. Thus, an increased proportion of Component X_1 is signified by moving to the left, along the base of the triangle. The essential point is that consistency of direction must be preserved along all three boundaries.

The interior of the triangle represents mixtures in which all three components are present. Point E, for example, represents a mixture of $0.3X_1$, $0.4X_2$, and $0.3X_3$. A line joining B and D represents all values of Component X_2, with Components X_1 and X_3 present in equal proportions. Point E falls on this line. In general terms, a line joining an apex to a given point on the opposite side of the triangle represents a constant ratio of two components, with an ever-decreasing proportion of the third. As stated above, the line AC represents a situation in which Component X_2 is absent. Lines denoting any other proportion of Component X_2 (e.g., $X_2=0.4$) are drawn

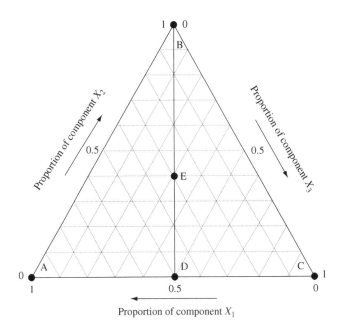

FIGURE 10.2 An equilateral triangle representing a three-component system.

parallel to AC. Thus, a line drawn parallel to one side of the triangle represents a constant proportion of one of the components.

Not all the area of the triangle may represent feasible formulations. Thus, if Component X_1 is the active ingredient in a tablet, then point C is impossible to achieve, because it would imply a zero content of active material. Similarly, a tablet containing only the drug (represented by point A) is, if not impossible, extremely unlikely.

Hence, in a formulation, there may be lower and upper limits to the proportions of a given component. Let us assume that lower limits are placed on the proportions of all three components, the limits being 0.20, 0.10, and 0.25 for X_1, X_2, and X_3, respectively. If these limits are transferred to Figure 10.2, the feasible space becomes the smaller equilateral triangle shown in Figure 10.3. Note that the imposition of lower limits on the components does not alter the shape of the figure. If the magnitudes of the lower limits are the same for all three components, then the resultant figure, as well as having the same shape as the original, has the same center point. Also note that all the three components cannot simultaneously assume their minimum values, because these would total 0.55, rather than unity, and hence would not form a valid mixture.

If lower boundaries are placed on the values of the components, then attention can be focused on to a subregion of the original space. It may then be useful to redefine the coordinates of the region in terms of "pseudocomponents." If these three pseudocomponents are represented by X'_1, X'_2, and X'_3, respectively, then each corner of the smaller triangle represents the situation where only one of the pseudocomponents is present, even though all three of the actual components are

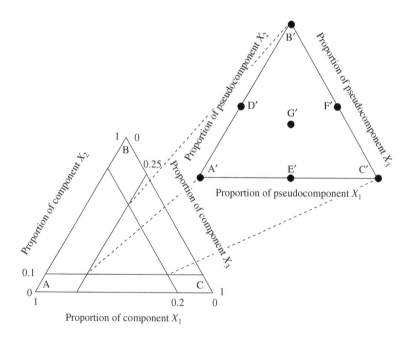

FIGURE 10.3 Triangular diagram representing a three-component system, all components having lower limits.

TABLE 10.1
Pseudocomponents and Actual Composition of Points A′ to G′ in Figure 10.3

Point	Pseudocomponent Composition			Actual Composition		
	X'_1	X'_2	X'_3	X_1	X_2	X_3
A′	1.00	0.00	0.00	0.65	0.10	0.25
B′	0.00	1.00	0.00	0.20	0.55	0.25
C′	0.00	0.00	1.00	0.20	0.10	0.70
D′	0.50	0.50	0.00	0.42	0.33	0.25
E′	0.50	0.00	0.50	0.42	0.10	0.48
F′	0.00	0.50	0.50	0.20	0.33	0.48
G′	0.33	0.33	0.33	0.35	0.25	0.40

present. Consider point A′. This represents Pseudocomponent X'_1 in a proportion of unity, yet the actual composition of this mixture is 0.65 of X_1, 0.10 of X_2, and 0.25 of X_3. The pseudocomponent and original component composition of several points in Figure 10.3 are shown in Table 10.1. The use of pseudocomponents and proportions based on them rather than on the actual proportions present is helpful in the computation of models. See by analogy the use of coded values for model-dependent optimization in Chapter 8.

If lower limits are introduced into the permissible ranges of components, then the shape of the resultant space does not change. If an upper limit, or both upper and lower limits are imposed, then the shape of the space is changed. The unhatched area in Figure 10.4, an irregular pentagon, represents a situation in which Component X_1 lies between 0.25 and 0.60, Component X_2 between 0.20 and 0.75, and Component X_3 between 0.10 and 0.35. The available space for the design, in addition to being a different shape to the original, is greatly restricted. All the components cannot be at their minimum or their maximum values at the same time, because the sum of the three proportions then would either be less than or exceed unity.

One of the components of the mixture may be in a considerable excess over all the other components, which in turn vary within relatively narrow limits. Examples of this could be the solvent in a liquid formulation, or a diluent in a tablet or capsule. In such cases, this component can be excluded, as its proportion will show little change. The other ingredients are then independent factors and can be treated by methods such as factorial design described in earlier chapters.

10.3 MIXTURES WITH MORE THAN THREE COMPONENTS

If there are four components (X_1, X_2, X_3, and X_4), all of which can vary in proportion from zero to unity, then there are three dimensions and four vertices, and the space is represented by a regular tetrahedron (Figure 10.5). As before, $X_1 + X_2 + X_3 + X_4 = 1$. If, however, the proportion of one of the components is fixed, then all mixtures of

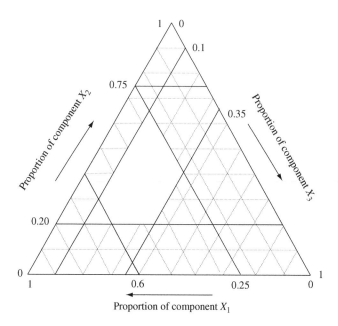

FIGURE 10.4 Triangular diagram representing a three-component system, all components having both upper and lower limits. The unhatched area represents the available experimental space.

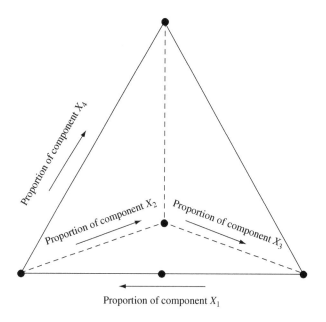

FIGURE 10.5 A regular tetrahedron representing a four-component system.

the other three can be represented by an equilateral triangle, using the method of pseudocomponents described earlier.

Components X_1, X_2, and X_3 form the triangular base of the tetrahedron. The fourth component X_4 is represented by measurement in a vertical direction away from the center of the base. Thus, the top point of the figure represents a "mixture" composed entirely of Component X_4. A point halfway between the top point and the center of the base represents a mixture, of which half is represented by Component X_4 and the other half by equal proportions of Components X_1, X_2, and X_3. Guidance on the construction of diagrams for four-component systems is given by Findlay.[1]

Four is the highest number of components that can be depicted diagrammatically, though with suitable mathematical techniques, the number of components that can be considered is unlimited.

10.4 RESPONSE-SURFACE METHODOLOGY IN EXPERIMENTS WITH MIXTURES

The properties of mixtures can be studied by response-surface methodology and mixtures can be optimized by using adaptations of the techniques described in Chapters 7 and 8. As before, this is best illustrated by a simple example.

The objective of the experiment is to study the solubility of Compound A in blends of three solvents, namely, ethanol, propylene glycol, and water. Thus, the "factor" in this experiment is the composition of the solvent blend and the response is the solubility of Compound A. The next stage is to decide whether there are any lower or upper limits to the proportions of each of the components. This will define the shape and size of the space available for the experiment design. The decision whether to consider the actual proportions of the components or of the pseudocomponents can now be taken. As all possible blends are acceptable in the current experiment, no limits are applied.

Then, the number of experiments and their position in the factor space must be considered. Obviously, for reasons of economy, the number of experiments should be as low as possible, but the decision will depend on whether the results are to be assessed by model-dependent or model-independent methods.

To maintain consistency with the methods used in Chapters 7 and 8, the proportions of the three components will be represented by the terms X_1, X_2, and X_3 in the subsequent discussion. If model-dependent methods are to be used, the number of experiments will depend on whether or not a linear relationship between the response and the composition of the mixture is anticipated.

10.4.1 Rectilinear Relationships between Composition and Response

If a linear relationship is expected, then three experimental points are chosen at the vertices of the triangle representing pure components only, and the response is measured. These three experiments are designated Mixtures 1, 2, and 3 in Figure 10.6 and in Table 10.2.

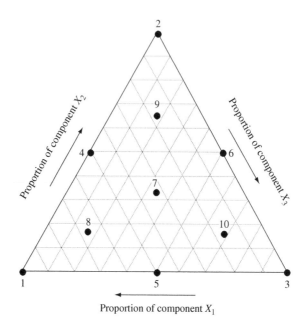

FIGURE 10.6 Experimental design for a three-component mixture assuming a linear relationship between composition and response.

TABLE 10.2
Composition of Mixtures of Ethanol, Propylene Glycol, and Water and the Solubility (g·l⁻¹) of Compound A in These Mixtures at 20°C

Mixture Number	Composition			Solubility of A	
	Ethanol (X_1)	Propylene Glycol (X_2)	Water (X_3)	Measured	Predicted by (10.4)
1	1.0	0.0	0.0	6.5	6.5
2	0.0	1.0	0.0	3.3	3.3
3	0.0	0.0	1.0	1.1	1.1
4	0.5	0.5	0.0	4.6	4.9.
5	0.5	0.0	0.5	2.6	3.8
6	0.0	0.5	0.5	2.3	2.2
7	0.33	0.33	0.33	1.7	3.6
8	0.67	0.17	0.17	3.6	
9	0.17	0.67	0.17	3.3	
10	0.17	0.17	0.67	1.3	

The data can be fitted to an equation of the form of (10.1)

$$Y = \beta_1 X_1 + \beta_2 X_2 + \beta_3 X_3 + e \tag{10.1}$$

where
Y = the response
β_1, β_2, and β_3 = the coefficients.

But $X_1 + X_2 + X_3 = 1$.

Therefore, $X_3 = 1 - (X_1 + X_2)$.

Substitution into (10.1) gives (10.2)

$$Y = \beta_1 X_1 + \beta_2 X_2 + \beta_3 - \beta_3 X_1 - \beta_3 X_2 \tag{10.2}$$

Rearrangement of (10.2) gives (10.3)

$$Y = (\beta_1 - \beta_3)X_1 + (\beta_2 - \beta_3)X_2 + \beta_3 \tag{10.3}$$

At the corner of the triangle where $X_1 = 1$, X_2 and X_3 must be 0. Therefore, by substitution of the value of the response at this corner into (10.1), the value of β_1 can be obtained. Coefficients β_2 and β_3 can be calculated in the same way.

Equation (10.1) thus becomes (10.4)

$$Y = 6.5X_1 + 3.3X_2 + 1.1X_3 \tag{10.4}$$

To establish the validity of the linear model equation, it is necessary to measure the responses at other points. Suitable blends are at the midpoints of each side and also at the central point of the triangle. The former (designated Mixtures 4, 5, and 6 in Table 10.2 and Figure 10.6) represent blends of equal proportions of two solvents, the third being absent. The latter (designated Mixture 7) represents a blend of equal proportions of all three solvents. The measured solubilities and those predicted by (10.4) are given in Table 10.2. The differences between observed and predicted solubilities are quite large, and a more complex relationship between solvent blend and solubility is therefore worth investigating.

If lower limits had been introduced to the proportions of some or all of the components, then similar positions on the restricted design space would be chosen. If lower and upper limits were imposed, the points would be dispersed across the irregularly shaped space.

10.4.2 DERIVATION OF CONTOUR PLOTS FROM RECTILINEAR MODELS

The regression equation can be used to derive contours linking combinations of solvents, giving equal solubilities of Compound A. The values of all three variables, X_1, X_2, and X_3, can be substituted into the equation. As the values of the coefficients are known, the responses can be calculated. If this is done at, for example, intervals representing a proportion of 0.1 for each component, the response at each point can be calculated and the position of the contour lines quickly, if only approximately, established.

Alternatively, (10.3) can be rearranged to give (10.5)

$$X_2 = \frac{Y - \beta_3 - (\beta_1 - \beta_3)X_1}{(\beta_2 - \beta_3)} \tag{10.5}$$

Thus, if the coefficients β_1, β_2, and β_3 are known, then for any given value of X_1, the value of X_2, which will give a specified response for Y, can be calculated.

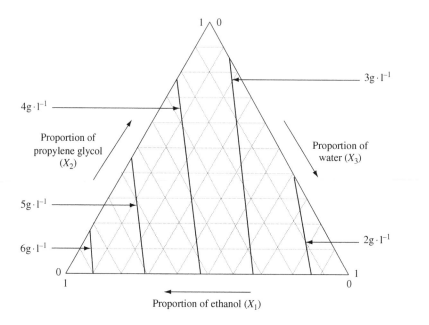

FIGURE 10.7 Contour plot of solubility of Compound A ($g \cdot l^{-1}$) in blends of three solvents, derived from (10.4).

A contour plot derived from (10.4) is shown in Figure 10.7. It consists of a series of straight lines.

10.4.3 HIGHER-ORDER RELATIONSHIPS BETWEEN COMPOSITION AND RESPONSE

If a higher-order relationship is suspected, then more experiments must be carried out because there will be more coefficients in the model equation. A second-order equation linking the response to the proportions of the three solvents is given by (10.6)

$$Y = \beta_1 X_1 + \beta_2 X_2 + \beta_3 X_3 + \beta_{12} X_1 X_2 + \beta_{13} X_1 X_3 + \beta_{23} X_2 X_3 + e \tag{10.6}$$

It should be noted that there are no squared terms in this equation because, for example, X_1^2 is replaced by $X_1(1 - X_2 - X_3)$. The products of this expression, namely, X_1, $X_1 X_2$, and $X_1 X_3$, are included in the first-order and interaction terms.

There are six coefficients in (10.6), so an experimental design using Mixtures 1 to 6 described in Table 10.2 would give a saturated design. It would be better to include at least one more experiment in the design, and Mixture 7 is the most suitable. The coefficients of (10.6) can be determined by multiple regression, as described in earlier chapters, the multiple regression program being modified so that there is no constant term (β_0). Equation (10.6) becomes (10.7), with a coefficient of determination of 0.9500

$$Y = 6.58X_1 + 3.38X_2 + 1.18X_3 - 2.94X_1X_2 - 6.54X_1X_3 - 1.34X_2X_3 \quad (10.7)$$

The equation can be validated by measuring the solubility at points not included in the original design. These are Mixtures 8, 9, and 10, and their composition is given in Table 10.3. The measured solubilities and those predicted by (10.7) are also given in Table 10.3.

The model can be developed further by including the three-way interaction term in the model equation, giving (10.8). This is known as a reduced cubic equation. Its derivation is beyond the scope of this chapter, but the interested reader is referred to Scheffé.[2]

$$Y = \beta_1X_1 + \beta_2X_2 + \beta_3X_3 + \beta_{12}X_1X_2 + \beta_{13}X_1X_3 + \beta_{23}X_2X_3 + \beta_{123}X_1X_2X_3 \quad (10.8)$$

There are now seven coefficients in the equation. Using Mixtures 1 to 7 will give a saturated design, so additional design points should be included. These are Mixtures 8 to 10 in Table 10.3.

Multiple regression gives (10.9), and the solubilities predicted by (10.9) are given in Table 10.4. The coefficient of determination is now 0.9840, indicating that inclusion of the $X_1X_2X_3$ term has improved the fit of the model to the data.

$$\begin{aligned} Y = {} & 6.42X_1 + 3.45X_2 + 1.09X_3 - 1.06X_1X_2 - 5.00X_1X_3 + 0.69X_2X_3 \\ & - 31.72X_1X_2X_3 \end{aligned} \quad (10.9)$$

Stephens et al.[3] have adopted this approach to formulate a solution of a vitamin D_2 analog.

TABLE 10.3
Composition of Mixtures of Ethanol, Propylene Glycol, and Water and the Solubility ($g \cdot l^{-1}$) of Compound A in These Mixtures at 20 °C

Mixture Number	Composition			Solubility of A	
	Ethanol (X_1)	Propylene Glycol (X_2)	Water (X_3)	Measured	Predicted by (10.7)
1	1.0	0.0	0.0	6.5	6.58
2	0.0	1.0	0.0	3.3	3.38
3	0.0	0.0	1.0	1.1	1.18
4	0.5	0.5	0.0	4.6	4.25
5	0.5	0.0	0.5	2.6	2.25
6	0.0	0.5	0.5	2.3	1.95
7	0.33	0.33	0.33	1.7	2.49
8	0.67	0.17	0.17	3.6	4.07
9	0.17	0.67	0.17	3.3	2.90
10	0.17	0.17	0.67	1.3	1.50

TABLE 10.4
Composition of Mixtures of Ethanol, Propylene Glycol, and Water and the Solubility (g·l^{-1}) of Compound A in These Mixtures at 20 °C

Mixture Number	Composition			Solubility of A	
	Ethanol (X_1)	Propylene Glycol (X_2)	Water (X_3)	Measured	Predicted by (10.9)
1	1.0	0.0	0.0	6.5	6.42
2	0.0	1.0	0.0	3.3	3.45
3	0.0	0.0	1.0	1.1	1.09
4	0.5	0.5	0.0	4.6	4.67
5	0.5	0.0	0.5	2.6	2.50
6	0.0	0.5	0.5	2.3	2.44
7	0.33	0.33	0.33	1.7	1.89
8	0.67	0.17	0.17	3.6	3.79
9	0.17	0.67	0.17	3.3	2.79
10	0.17	0.17	0.67	1.3	1.27

10.4.4 CONTOUR PLOTS DERIVED FROM HIGHER-ORDER EQUATIONS

Substituting $X_3 = (1 - X_1 - X_2)$ into (10.6) gives (10.10)

$$Y = \beta_1 X_1 + \beta_2 X_2 + \beta_3 (1 - X_1 - X_2) + \beta_{12} X_1 X_2 + \beta_{13} X_1 (1 - X_1 - X_2)$$
$$+ \beta_{23} X_2 (1 - X_1 - X_2) \tag{10.10}$$

Multiplying out and gathering the terms to give a quadratic equation in X_2 yields (10.11)

$$Y = -\beta_{23} X_2^2 + (\beta_2 - \beta_3 + \beta_{12} X_1 - \beta_{13} X_1 + \beta_{23} - \beta_{23} X_1) X_2 +$$
$$(\beta_1 X_1 + \beta_3 - \beta_3 X_1 + \beta_{13} X_1 - \beta_{13} X_1^2 - Y) \tag{10.11}$$

The general solution of a quadratic equation is given in (10.12)

$$X = \frac{-b \pm \sqrt{b^2 - 4ac}}{2a} \tag{10.12}$$

In this case, a, b, and c, the coefficients of the X^2, X^1, and X^0 terms, respectively, are as follows:

$$a = -\beta_{23}$$
$$b = (\beta_2 - \beta_3 + \beta_{12} X_1 - \beta_{13} X_1 + \beta_{23} - \beta_{23} X_1)$$
$$c = (\beta_1 X_1 + \beta_3 - \beta_3 X_1 + \beta_{13} X_1 - \beta_{13} X_1^2 - Y)$$

The coefficients are known from the regression calculation, and Y is the required response that is the value of the contour. Hence, if a value of X_1 is selected, the corresponding values of X_2 can be calculated. Note that there will be two values of X_2. These are the roots of the equation that satisfy (10.11). Some of these roots will not be applicable in problems of this type. For example, the value of X_2 cannot be negative. Neither can values of X_2 be accepted which, when added to the corresponding value of X_1, give a sum in excess of unity, because this would result in an impossible negative value for X_3. If two inappropriate values for X_2 are obtained, this means that for the chosen value of Y, a solution of the equation within the designated space is not possible.

If a reduced cubic equation (10.8) is used as the model, this too is transformed into a quadratic equation. In this case, the coefficients for the X^2, X^1, and X^0 terms are:

$$a = -(\beta_{23} + \beta_{123}X_1)$$
$$b = (\beta_2 - \beta_3 + \beta_{12}X_1 - \beta_{13}X_1 + \beta_{23} + \beta_{23}X_1 + \beta_{123}X_1 - \beta_{123}X_1^2)$$
$$c = (\beta_1X_1 + \beta_3 - \beta_3X_1 + \beta_{13}X_1 - \beta_{13}X_1^2)$$

Contours derived from second-order model equations and reduced cubic equations are curves. Figure 10.8 shows contours of solubility of Compound A in blends of three solvents obtained from (10.9).

A full discussion of the selection of appropriate designs and models is given by Huisman et al.[4]

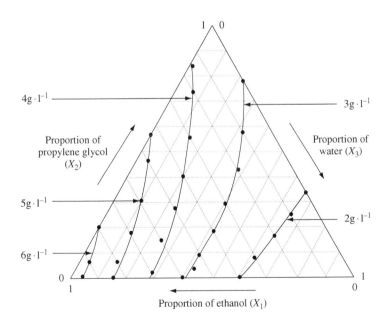

FIGURE 10.8 Contour plot of solubility of Compound A ($g \cdot l^{-1}$) in blends of three solvents, derived from (10.9).

10.5 THE OPTIMIZATION OF MIXTURES

The application of optimization techniques to three-component mixtures can be illustrated by an extension of the example used earlier in this chapter.

Compound A is to be dissolved in a mixture of three liquids: ethanol, propylene glycol, and water. It is anticipated that many blends of these three liquids will provide a satisfactory solvent system, but the cheapest possible mixture should be identified. Thus, the problem can be divided into two parts: the question of solubility and the question of cost. The latter, being more straightforward, will be addressed first.

It is reasonable to expect that the cost of a mixture of liquids is directly related to the proportion of each component in that mixture. Thus, if C is the cost of the mixture, then

$$C = \beta_1 X_1 + \beta_2 X_2 + \beta_3 X_3 \tag{10.13}$$

where

X_1, X_2, and X_3=the proportions of ethanol, propylene glycol, and water, respectively
β_1, β_2, and β_3=their respective coefficients.

Let us assume that the costs per liter of the three liquids are £10.00 for ethanol, £5.00 for propylene glycol, and £0.50 for water.

The coefficients in (10.13) are easily calculated. Imagine a "mixture" containing only ethanol (X_1). Then the cost of one liter of this "mixture" will be the same as that of one liter of ethanol, that is, £10.00. It follows that the coefficient $\beta_1 = 10.00$. Using similar arguments, $\beta_2 = 5.00$ and $\beta_3 = 0.50$.

Equation (10.13) now becomes (10.14)

$$C = 10.00X_1 + 5.00X_2 + 0.50X_3 \tag{10.14}$$

Thus, the cost of any blend of the three solvents can be calculated from this equation. For example, a mixture containing proportions of 0.5 ethanol, 0.3 propylene glycol, and 0.2 water will cost (£10.00×0.5)+(£5.00×0.3)+(£0.50×0.2)=£6.60.

The factor space for this equation can be represented by the triangle shown in Figure 10.9. Because there are only three terms in (10.14), and a linear relationship can be assumed to apply between the total cost and the proportions of each liquid, then an accurate model for the whole factor space can be derived from only three data points. These are the apices of the triangle, points 1, 2, and 3 on Figure 10.9.

A contour plot for costs can be derived from (10.14), which now becomes (10.15)

$$(0.50 - 5.00)X_2 = (10.00 - 0.50)X_1 + 0.50 - C \tag{10.15}$$

If, for example, the £6.00 contour is required, and hence 6.00 is substituted for C in (10.15), then for specified values of X_1, the corresponding value of X_2 can be calculated. Thus, if X_1 (the proportion of ethanol)=0.5, then a proportion of 0.167 propylene glycol is needed. The remainder (0.333) is water, and the total cost of

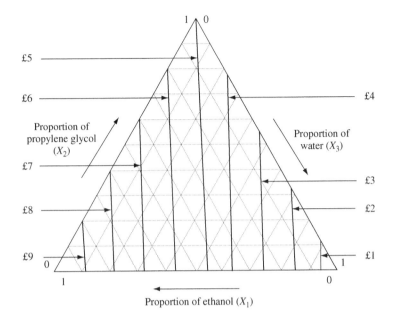

FIGURE 10.9 Contour plot of the cost of mixtures of ethanol, propylene glycol, and water, derived from (10.15).

the mixture is £6.00. Contours of the different total costs are shown in Figure 10.9. Note that they are straight lines.

The question of the solubility of Compound A can now be addressed. It is uncertain whether a simple proportionality will apply to solubility data in three-component solvents, and it is therefore prudent to use a higher-order model equation with interaction terms. The reduced cubic model (10.9) has already been shown to give a good representation of the solubility data, which is given in Table 10.3.

The possibility of optimal solutions to this problem should now be apparent. Though water is by far the cheapest solvent, the solubility of Compound A in water is lower than in the other two liquids, and the best solvent (ethanol) is the most expensive.

The contour plot of the cost data (Figure 10.9) can now be superimposed on the solubility contour plot (Figure 10.8), giving Figure 10.10.

In many experiments involving solubility, the required concentration of the solute will be known, because the usual objective is to prepare a solution containing a specified weight of solute in a given volume of solution. For example, in this case, let the required concentration be $3 \, g \cdot l^{-1}$. The cheapest blend of solvents that will give the required solubility for Compound A can now be read off from Figure 10.10.

10.6 PARETO-OPTIMALITY AND MIXTURES

The above example of optimizing cost and solubility can also be approached using Pareto-optimality techniques. The two regression equations (10.14) and (10.9) are

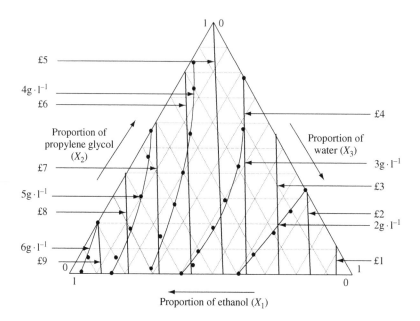

FIGURE 10.10 Combined contour plots of the solubility of Compound A in mixtures of ethanol, propylene glycol, and water, and the cost of those mixtures.

obtained and the goodness of fit determined as before. Convenient values of the proportions of the three liquids are now chosen, for example, by selecting values at intervals of 0.1. Thus, $X_1=0.0$, $X_2=0.0$, $X_3=1.0$, followed by $X_1=0.0$, $X_2=0.1$, $X_3=0.9$, and so forth. There are 66 such points. Substituting these combinations and the coefficients into the two regression equations gives the responses. Thus, for example, for the point representing $X_1=0.8$, $X_2=0.1$, $X_3=0.1$, the cost is £8.55 and the solubility is $4.9\,g\cdot l^{-1}$.

By carrying out this calculation for all 66 points, 66 pairs of cost and solubility data are obtained. These are shown in Figure 10.11, point Z representing the two items of information for $X_1=0.8$, $X_2=0.1$, $X_3=0.1$.

Any point in Figure 10.11, for example, point P, is selected, and two intersecting and perpendicular lines are then drawn through this point, as described in Chapter 8, dividing the space into four quadrants. Because the cost is to be minimized, and the solubility maximized, Quadrant II is the quadrant of interest. The Pareto-optimal points are shown in Figure 10.11, and the corresponding compositions, cost, and solvent power are given in Table 10.5.

For all these points, a higher solubility cannot be obtained without an increase in cost. If a specific solubility is required (e.g., $3\,g\cdot l^{-1}$), then the lowest cost of a mixture which can produce such a solubility is given by the point of intersection of a line drawn at $3\,g\cdot l^{-1}$ with the line joining the Pareto-optimal points. This occurs at about £4.00 per liter.

The application of Pareto-optimality to mixture designs with a special reference to the selection of solvents for high-performance liquid chromatography (HPLC)

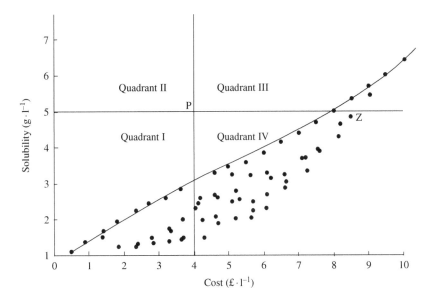

FIGURE 10.11 Pareto-optimal diagram of the cost and solvent power of mixtures of ethanol, propylene glycol, and water.

has been described by Smilde et al.[5] The technique has been used in tablet formulation by De Boer et al.[6]

10.7 PROCESS VARIABLES IN MIXTURE EXPERIMENTS

The earlier discussion in this chapter has dealt entirely with mixtures in which the composition of the mixture was changed. However, in addition to being affected by the composition of the mixture, the response may be affected by process factors or environmental conditions.

In general terms, suppose that there is a mixture of q components. There are n process variables, which are to be studied at two levels: $z=-1$ and $z=+1$. Such a design could be the three-component liquid mixtures referred to earlier in this chapter, with the solubilities measured at a temperature other than 20 °C. In this, there are three components ($q=3$) and one process variable, the temperature. Hence, $n=1$. The experimental design at each value of the process variable is represented by an equilateral triangle, using seven mixtures of solvents and solubility being measured at 20 °C as before. Thus, the value of $z=-1$ is 20 °C. All the experiments are then repeated at a higher temperature (say, 40 °C). The experimental design is shown in Figure 10.12 and consists of two equilateral triangles, one representing mixtures of the three components at 20 °C and the other the same mixtures at 40 °C. Point A in this diagram represents a solvent system comprising equal proportions of the three components at 40 °C. The corresponding mixture at 20 °C is represented by point B.

TABLE 10.5
Pareto-Optimal Points Obtained from Figure 10.11: Their Composition and Resultant Cost ($£ \cdot l^{-1}$) and Solvent Power ($g \cdot l^{-1}$)

Cost ($£ \cdot l^{-1}$)	Solvent Power ($g \cdot l^{-1}$)	Proportion of Ethanol (X_1)	Proportion of Propylene Glycol (X_2)	Proportion of Water (X_3)
0.50	1.09	0.0	0.0	1.0
0.95	1.39	0.0	0.1	0.9
1.40	1.67	0.0	0.2	0.8
1.85	1.94	0.0	0.3	0.7
2.30	2.20	0.0	0.4	0.6
2.75	2.44	0.0	0.5	0.5
3.20	2.67	0.0	0.6	0.4
3.65	2.89	0.0	0.7	0.3
4.10	3.09	0.0	0.8	0.2
4.55	3.28	0.0	0.9	0.1
5.00	3.45	0.0	1.0	0.0
5.50	3.65	0.1	0.9	0.0
6.00	3.87	0.2	0.8	0.0
6.50	4.12	0.3	0.7	0.0
7.00	4.38	0.4	0.6	0.0
7.50	4.67	0.5	0.5	0.0
8.00	4.98	0.6	0.4	0.0
8.50	5.31	0.7	0.3	0.0
9.00	5.63	0.8	0.2	0.0
9.50	6.03	0.9	0.1	0.0
10.00	6.42	1.0	0.0	0.0

The data is derived from Figure 10.11.

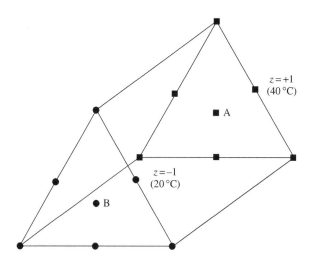

FIGURE 10.12 Combined experimental design for a three-component mixture and one process variable studied at two levels.

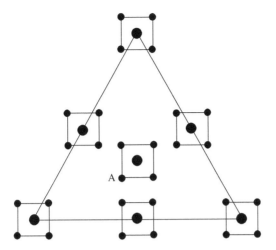

FIGURE 10.13 Combined experimental design for a three-component mixture with two process variables studied at two levels.

More elaborate factorials can also be incorporated into each point of the mixture diagram. Thus, if there are two variables of interest (n_1 and n_2) to be studied at two levels, then each mixture is studied at four combinations of the two factors. This design can be represented as a square that is repeated for every mixture. Figure 10.13 shows such a design for seven mixtures of liquids. In this case, point A represents a mixture of equal proportions of three components, with both factors at their lower level.

For a full discussion of the design of such experiments and the mathematical treatment of the data, the reader is referred to Cornell.[7] Duineveld et al.[8] have applied designs incorporating mixture and process variables to tablet formulation.

FURTHER READING

The following articles describe the use of mixture designs in the design and evaluation of experiments. Reference is made to two review articles, followed by a selected bibliography.

De Boer, J. H., Smilde, A. G., and Doornbos, D. A., Introduction of multi-criteria decision making in optimization procedures for pharmaceutical formulations, *Acta Pharm. Technol.*, 34, 140, 1988.

Lewis, G. A., Optimization methods, in *Encyclopaedia of Pharmaceutical Technology*, Swarbrick, J. and Boylan, J. C., Eds., Dekker, New York, vol. 2, 2002, p. 1922.

Anik, S. T. and Sukumar, L., Extreme vertexes design in formulation development: solubility of butoconazole nitrate in a multicomponent system, *J. Pharm. Sci.*, 70, 897, 1981.

Chu, J. S. et al., Mixture experimental design in the development of a mucoadhesive gel formulation, *Pharm. Res.*, 8, 1401, 1991.

Geoffroy, J. M., Fredrickson, J. K., and Shelton, J. T., A mixture experiment approach for controlling the dissolution rate of a sustained-release product, *Drug Dev. Ind. Pharm.*, 24, 799, 1998.

Hariharan, M. et al., Effect of formulation composition on the properties of controlled release tablets prepared by roller compaction, *Drug Dev. Ind. Pharm.*, 30, 565, 2004.

Konkel, K. and Mielck, J. B., A compaction study of directly compressible vitamin preparations for the development of a chewable tablet: Part I, *Pharm. Technol.*, March, 138, 1992.

Marti-Mestres, G. et al., Optimisation with experimental design of nonionic, anionic and amphoteric surfactants in a mixed system, *Drug Dev. Ind. Pharm.*, 23, 993, 1997.

Minarro, M. et al., Study of formulation parameters by factorial design in metoprolol tartrate matrix systems, *Drug Dev. Ind. Pharm.*, 27, 965, 2001.

Ramachandran, S., Chen, S., and Etzler, F., Rheological characterization of hydroxypropyl cellulose gels, *Drug Dev. Ind. Pharm.*, 25, 153, 1999.

Rambali, B. et al., Itraconazole formulation studies of the melt extrusion process with mixture designs, *Drug Dev. Ind. Pharm.*, 29, 641, 2003.

Van Kamp, H. V., Bolhuis, G. K., and Lerk, C. F., Optimization of a formulation based on lactoses for direct compression, *Acta Pharm. Technol.*, 34, 11, 1988.

Vojnovic, D. and Chicco, D., Mixture experimental designs applied to solubility predictions, *Drug Dev. Ind. Pharm.*, 23, 639, 1997.

Yazici, E. et al., Phenytoin microcapsules: bench scale formulation, process characterisation and release kinetics, *Pharm. Dev. Technol.*, 1, 175, 1996.

REFERENCES

1. Findlay, A., *The Phase Rule and its Applications*, 9th ed., Dover, New York, 1951.
2. Scheffé, H., Experiments with mixtures, *J. R. Stat. Soc., Ser. B*, 20, 344, 1958.
3. Stephens, D. et al., A statistical experimental approach to co-solvent formulation of a water soluble drug, *Drug Dev. Ind. Pharm.*, 25, 961, 1999.
4. Huisman, R. et al., Development and optimization of pharmaceutical formulations using a simplex lattice design, *Pharmaceutisch Weekblad. Scientific Edition*, 6, 185, 1984.
5. Smilde, A. K., Knevelman, A., and Coenegracht, P. M. J., Introduction of multi-criteria decision making in optimization procedures for high-performance liquid chromatographic separations, *J. Chromatogr.*, 369, 1, 1986.
6. De Boer, J. H., Bolhuis, G. K., and Doornbos, D. A., Comparative evaluation of multi-criteria decision making and combined contour plots in optimization of directly compressed tablets, *Eur. J. Pharm. Biopharm.*, 37, 159, 1991.
7. Cornell, J. A., *Experiments with Mixtures*, 3rd ed., Wiley, New York, 2002.
8. Duineveld, C. A. A., Smilde, A. K., and Doornbos, D. A., Designs for mixture and process variables applied in tablet formulations, *Anal. Chim. Acta*, 277, 455, 1993.

11 Artificial Neural Networks and Experimental Design

11.1 INTRODUCTION

Much of the pharmaceutical development process involves the optimization of formulation and process variables, as described in Chapters 7 and 8. Prediction of the behavior of formulations is often difficult, and response-surface methodology has proven useful in this respect. In model-dependent optimization, the regression line which best fits the independent variables (formulation or processing factors) and the responses of a series of experiments is determined. The key point is that the type of regression line—rectilinear or quadratic, with or without interaction terms—is preselected by the experimenter, together with predetermined statistical significance levels. Indeed, the chosen experimental design will at least in part be governed by the relationship into which the data are to be fitted. For example, a design of at least three levels is essential if a quadratic relationship is to be explored.

The fitting of data to a regression line is now almost invariably carried out by the use of a computer. This essentially consists of an arithmetic/logic unit that manipulates data held in memory. Instructions control where the data are held, what is done with them, and where the result is to be stored. Such instructions are termed algorithms, which can be defined as a series of steps that achieve a desired aim. Computers work because of hierarchies of and interactions between algorithms, all of which must be devised by human beings. The range of tasks which can be carried out by a conventional computer is limited to those for which algorithms can be devised, and every step of the process must be spelt out to it.

In recent years, considerable use has been made in pharmaceutical studies of expert systems that attempt to capture the knowledge and experience of experts in a defined area.[1,2] These too are algorithmic methods, in that factual information is supplied together with rules for using that information. Hence, such systems are not creative and only deal with situations which have been anticipated. Nevertheless, expert systems have brought benefits such as improved knowledge availability and protection and consistency in the use of such information as well as cost savings. Indeed, Rowe[1] has claimed that with an efficient expert system, the optimization process, as described in earlier chapters of this book, is redundant.

The introduction of artificial neural networks (ANNs) (neural computing, parallel distributed processing, and connectionism are all synonyms) offers an alternative approach. Neural computing has been defined by Aleksander and Morton[3] as "the study of networks of adaptable nodes which, through a process of learning from task examples, store experiential knowledge and make it available for use." This definition is wide enough to include the living brain (hence the use of the term neural) which acquires knowledge by being exposed to examples, that is, we learn by experience. ANNs are therefore attempts to create machines that learn from experience in a similar way to the brain, though since little is known of the detail of how the brain actually works, the analogy should not be pressed too strongly.

The underlying function of a neural network is to identify patterns, that is, when presented with an input pattern, it produces an output pattern. To do this, the network has to recognize the relationship between input data and the corresponding response. Thus, in contrast to response-surface methodology, it is the neural network that identifies the relationship rather than the experimenter *a priori* deciding what form the relationship is to take and then fitting the input and output to that relationship.

However, there is more to neural networks than just being able to recognize patterns. Neural networks have the ability to learn and to generalize. Learning is achieved during the training phase, in which the network gains experience as it attempts to learn the underlying relationships between input and output. Having established these relationships, the network can then apply them to input data not previously encountered in the training phase and predict the outcome.

ANNs consist of input devices, interconnected processing elements or nodes, and output devices. Typically, there is one input layer, one output layer, and one or more hidden layers between them. Figure 11.1 shows a network with three inputs (the independent variables), a hidden layer with several processing elements, and two outputs (the responses). The connections between the individual units represent the architecture of the system. Processing takes place in the hidden layer(s) and the output layer but not in the input layer.

Consider the processing unit Q in the hidden layer and the three units P_1, P_2, and P_3 in the input layer. P_1, P_2, and P_3 have outputs O_1, O_2, and O_3, respectively.

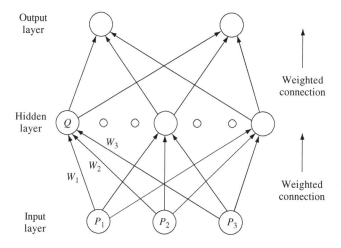

FIGURE 11.1 A fully interconnected artificial neural network with an input layer, one hidden layer and an output layer.

The strength of the connection between any two units is called the weight and, for the connections among P_1, P_2, and P_3 and Q, are W_1, W_2, and W_3, respectively. The three outputs to unit Q are summed (S_Q) to give (11.1)

$$S_Q = O_1 W_1 + O_2 W_2 + O_3 W_3 \qquad (11.1)$$

After S_Q is computed, it is transformed by the sigmoid transfer function (11.2) that gives an output between 0 and 1, both extreme values being approached asymptotically.

$$F_{S_Q} = \frac{1}{1 + e^{-S_Q}} \qquad (11.2)$$

This value is then transmitted to the next layer in the network, modified by the strength of the relevant connections, transformed again, and so on until the output layer is reached.

The network must now be trained, the purpose of training being to find a set of weight values that minimizes the differences between the output of the network and the measured value of the response. Sets of input data (the independent variables) and the corresponding responses are selected. Because the independent variables and the responses will almost certainly have different units and magnitudes, they must first be normalized, a process already encountered in Chapter 9. If L and U are the minimum and maximum values of a particular input or output, conversion of each value to a range between 0 and 1 is carried out using (11.3)

$$\text{converted value} = \frac{\text{value} - L}{U - L} \qquad (11.3)$$

The weighting of each connection is initially set at a low randomly chosen value. Input data are then fed into the network, modified by the weightings, and received by the output layer. The computed value of a particular output is compared with the known value of that response, and then by a process known as backpropagation, the weights are changed and the process is repeated. Weight changes are governed by one of many learning algorithms, the most commonly used being the generalized delta rule.

Training is thus an iterative process which continues either for a specified number of cycles or until the difference between the calculated value of the response and the observed value reaches a predefined level. In fact, the root mean square error is usually used as a measure of learning and reflects the degree to which the network has been trained.

The next stage is to feed other sets of values of the independent variables into the network. These data will be changed according to the weightings derived during the training phase, and the output of the network is then compared with the responses obtained experimentally. If the network is properly trained to a point where generalization is possible, then the predicted values and the observed values of the responses should be close to each other. If not, then the training process is unsatisfactory.

The network is not fitting the input and output data into a specific relationship. Thus, the question 'how well do the data fit the relationship?' cannot arise, and there is no objective test of fit such as the coefficient of determination (r^2). The computed relationship between input and output depends entirely on the training of the network. It is therefore essential that the set of training data cover the entire expected range of inputs, and "extrapolation," that is, using inputs from outside the training range, is particularly dangerous. Nor in the absence of a specific relationship between input and output is it possible to create a response-surface or contour plots.

Description of the many forms of architecture of ANNs and the rules governing weight and learning rate adjustment is beyond the scope of this book, but interested readers are referred to books such as *An Introduction to Neural Computing* by Aleksander and Morton,[3] *Neural Networks* by Picton,[4] and *Artificial Intelligence: a Modern Approach* by Russell and Norvig.[5] The number of nodes in the hidden layer is critical. If there are too few, then the network cannot detect patterns. If there are too many, then the network "memorizes" the patterns and its ability to generalize is handicapped. It has been suggested that if there are N input nodes, there should be at least $2N-1$ hidden nodes. This is known as the Kolmogorov theorem.[5]

11.1.1 PHARMACEUTICAL APPLICATIONS OF ANNs

Neural networks have been used in a wide variety of scientific purposes, because any problem that involves the processing of information from sensors to give a response should be amenable to the application of neural network technology. Thus, areas such as image processing, speech recognition, and fault diagnosis have all been examined using neural networks.

Achanta et al.[6] have suggested several areas in the medical and pharmaceutical sciences where neural networks might be useful. These include epidemiology, medical decision-making, and drug interactions. Until relatively recently, there has been little attempt to use neural networks in pharmaceutical development and technology. However, the possibility of such applications is now being realized.

One of the earliest applications of neural networking to pharmaceutical technology was made by Murtoniemi et al.,[7] who used it to model the fluidized bed granulation process.

The independent variables investigated by Murtoniemi et al. were inlet air temperature (40 °C, 50 °C, and 60 °C), atomizing air pressure (1.0 bar, 1.5 bar, and 2.0 bar), and the amount of binder solution (150 g, 300 g, and 450 g). Responses were granule size and friability. The original experiments were carried out by Merrku and Yliruusi,[8] and full experimental details can be found in their article.

The experimental design was a complete three-factor, three-level design, that is, there were 27 sets of experimental conditions. In addition, there were 11 replicated batches, so 38 batches of granulate were produced and tested. These batches constituted the training data for the network, and several network architectures were studied. There were three input processing units, because three experimental factors were being investigated, and there were two output units because two granule properties were of interest.

TABLE 11.1
Input Data (Factor Levels) and Output Data (Responses) to Test the Ability of the Artificial Neural Network to Generalize (T, Inlet Air Temperature; p, Atomizing Air Pressure; m, Binder Solution Amount)

Experiment Number	T (°C)	P (bar)	m (g)	Granule Size (μm)	Friability (%)
1	45	1.8	225	409	42.3
2	55	1.8	225	396	43.3
3	45	1.3	375	530	12.8
4	50	1.4	200	344	32.7
5	50	1.7	410	403	25.4

From Murtoniemi, E. et al., *Int. J. Pharm.*, 108, 15, 1994. With permission.

After training, the ability of the network to generalize was then tested by preparing five additional batches of granulate (Table 11.1). The values of the independent variables are all within the ranges used in the training data. There was thus no extrapolation.

An interesting feature of the work of Murtoniemi et al. was that the ability of the network to generalize was examined by increasing the number of processing units in the hidden layer and also by arranging these as one or two hidden layers. As there are three input parameters, the Kolmogorov theorem predicts that at least five hidden neurons should be present. After each change, the network was retrained and tested. After the addition of the 15th hidden neuron, no further improvement in the generalization ability of the network was achieved. In fact, the improvement brought about by increasing the number of units beyond 6 was very small, and no advantage accrued by having the hidden units in two rather than in one layer.

In an earlier part of this work, Merrku and Yliruusi[8] had derived regression equations relating the responses of granule size and friability to the three input factors. These equations could be used to predict granule size and friability for the five batches used to test the generalization of the network. Hence, a means was available to compare the predictive abilities of the neural network and regression equations. The results for granule size are shown in Figure 11.2.

In general, both the neural network and the regression equations led to a considerable underestimate of granule size, though the prediction by the network was closer to the experimental figures. A similar result was found with predictions for granule friability, though the underestimate was not always present in this case. A suggestion put forward by the authors for these underestimates was that the regression equations only included terms significant at $p < 0.05$. Though they were not statistically significant, the missing terms would have made some contribution to the response. Other studies in which the predictive properties of ANNs and conventional multivariate and regression methods are compared are given in the bibliography.

Chen et al.[9] compared the ability of four commercially available ANN programs to predict *in vitro* dissolution from controlled-release tablets. The input data

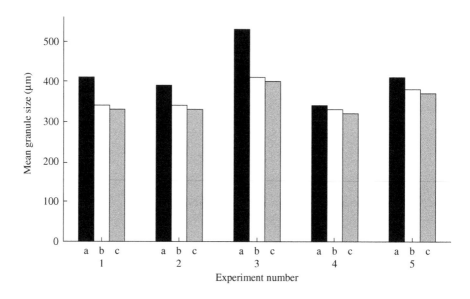

FIGURE 11.2   Mean granule size (μm): (a) experimental results, (b) results predicted by an artificial neural network, (c) results predicted by regression techniques (From Murtoniemi, E. et al., *Int. J. Pharm.*, 108, 15, 1994. With permission.)

consisted of ten variables, namely, seven independent formulation variables and three tablet variables—water content, granule size, and tablet hardness. Twenty-two different formulations were used. The output data were dissolution time profiles at ten sampling times. The optimum network architecture was obtained for each of the four systems by varying the number of hidden layers and the number of nodes in each layer. Interestingly, the optimum architectures were different, although the same data set was used in each case.

The four systems used and their optimal architectures were as follows:

- BrainMaker© v 3.7 (California Scientific Software, Nevada City, CA, U.S.A.), four nodes in the first hidden layer and eight nodes in the second hidden layer.
- CAD/CHEM© v 5.0 (AI Ware, Cleveland, OH, U.S.A.), nine nodes in a single hidden layer.
- NeuralWorks Professional II/PLUS© (NeuralWare, Pittsburgh, PA, U.S.A.), ten nodes in a single hidden layer.
- NeuroShell 2© v 3.0 (Ward Systems, Frederick, MD, U.S.A.), 8 nodes in the first hidden layer and 15 nodes in the second hidden layer.

Chen et al. plotted the observed percent dissolved data against the results predicted by the neural networks. If prediction had been perfect, each graph would have had a slope of +1 and a value of r^2 of 1.00. In fact, slopes ranged from 0.95 to 1.01, with an r^2 range of 0.95–0.99. The conclusion was drawn that all four programs gave reasonable predictions, with NeuroShell 2 as the best overall. The optimal architecture derived from this system is the only one that meets the requirements of the Kolmogorov theorem.

A study reported by Plumb et al.[10] is of particular interest. These workers studied the effect of six independent variables in a film coating formulation, the two responses of interest being crack velocity and film opacity. They found that the predictive ability of the ANN was superior to that of classical experimental designs, because some of the response surfaces were highly curved and were thus poorly represented by model equations. They suggested that classical experimental designs such as Box–Behnken and central composite designs were inappropriate for use in modeling by ANNs. Pseudorandom designs that covered the whole internal volume of the design space gave a much better prediction.

Bourquin et al.[11] also compared the predictive properties of ANNs and classical statistical methods. Using a tablet formulation with six independent variables and two responses, they too found that highly nonlinear relationships were difficult to model with classical designs and that the predictive properties of an ANN were much superior. However, they showed a potential weakness of the neural network method by introducing an outlying result into their data. This result was not detected by the network, and hence poor prediction was obtained. The regression equation, on the other hand, had a low value for the correlation coefficient, and this served as a warning that some of the data were suspect.

In addition to predicting responses, ANNs have also been used to determine optimal solutions. Several responses are incorporated into a single function to consider all those responses simultaneously. The desirability function of Derringer and Suich,[12] as described in Chapter 8, can be used for such purposes. Each individual output of the network can be normalized to partial desirability functions $d_1, d_2, d_3, \ldots, d_n$, which have values in the range 0–1. These are then combined into an overall desirability function D as shown in (11.4)

$$D = (d_1 \times d_2 \times d_3 \times \cdots \times d_n)^{1/n} \qquad (11.4)$$

Takayama et al.[13] have described such an approach to the formulation of a transdermal product, comparing their results with those obtained by classical methods.

FURTHER READING

Bourquin, J. et al., Advantages of artificial neural networks (ANNs) as alternative modelling technique for data showing non-linear relationships using data from a galenical study on a solid dosage form, *Eur. J. Pharm. Sci.*, 7, 5, 1998.

Chen, Y. et al., Prediction of drug content and hardness of intact tablets using artificial neural networks and near-infrared spectroscopy, *Drug Dev. Ind. Pharm.*, 27, 623, 2001.

Degim, T. et al., Prediction of skin penetration using artificial neural network (ANN) modelling, *J. Pharm. Sci.*, 92, 656, 2003.

Erb, R. J., Introduction of backpropagating neural network computation, *Pharm. Res.*, 10, 165, 1993.

Fu, X. C. et al., Predicting blood–brain barrier penetration of drugs using an artificial neural network, *Pharmazie*, 59, 126, 2004.

Hussain, A. S., Yu, X., and Johnson, R. D., Application of neural computing in pharmaceutical product development, *Pharm. Res.*, 8, 1248, 1991.

Kuppuswamy, R. et al., Practical limitations of tabletting indices, *Pharm. Dev. Technol.*, 6, 505, 2001.

Pillay, V. and Danckwearts, M. P., Textual profiling and statistical optimisation of cross-linked calcium alginate pertinate cellular acetophthalate gelisphere masses, *J. Pharm. Sci.*, 91, 2559, 2002.

Plumb, A. P. et al., Effect of varying optimisation parameters on optimisation by guided evolutionary simulated annealing (GESA) using a tablet film coat as an example formulation, *Eur. J. Pharm. Sci.*, 18, 259, 2003.

Reis, M. A. A., Sinisterra, R. D., and Belchor, J. C., An alternative approach based on artificial neural networks to study controlled release, *J. Pharm. Sci.*, 93, 418, 2004.

Rocksloh, K. et al., Optimisation of crushing strength and disintegration time of a high dose plant extract by neural networks, *Drug Dev. Ind. Pharm.*, 25, 1015, 1999.

Sathe, P. M. and Venitz, J., Comparison of neural networks and multiple linear regression as dissolution predictors, *Drug Dev. Ind. Pharm.*, 29, 349, 2003.

Takahara, J. et al., Multiobjective simultaneous optimisation based on artificial neural network in a ketoprofen hydrogel formula containing *O*-ethylmenthol as a percutaneous absorption enhancer, *Int. J. Pharm.*, 158, 203, 1997.

Takayama, K. et al., Neural network based optimisation of drug formulations, *Adv. Drug Del. Rev.*, 55, 1217, 2003.

Turkoglu, M., Ozarslan, R., and Sakr, A., Artificial neural network analysis for a direct compression tabletting study, *Eur. J. Pharm. Biopharm.*, 41, 315, 1995.

Wang, X. et al., Preparation and evaluation of high drug content matrices, *Drug Dev. Ind. Pharm.*, 29, 1109, 2003.

Wu, T. et al., Formulation optimisation technique based on an artificial neural network in salbutamol sulfate osmotic pump tablets, *Drug Dev. Ind. Pharm.*, 26, 211, 2000.

REFERENCES

1. Rowe, R. C., Expert systems, in *Encyclopaedia of Pharmaceutical Technology*, 2nd ed., Swarbrick, J. and Boylan, J. C., Eds., Marcel Dekker, New York, vol. 2, 2002, p. 1188.

2. Lai, S., Podczeck, F., Newton, J. M., and Daumesnil, R., Expert system to aid the development of capsule formulations. *Pharm. Technol. Eur.*, 8, 60, 62–64, 66, 68, 1996.

3. Aleksander, I. and Morton, H., *An Introduction to Neural Computing*, 2nd ed., Chapman & Hall, London, 1995.

4. Picton, P., *Neural Networks*, 2nd ed., Palgrave, London, 2000.

5. Russell, S. and Norvig, P., *Artificial Intelligence: a Modern Approach*, 2nd ed., Prentice-Hall, London, 2003.

6. Achanta, A. S., Kowalski, J. G., and Rhodes, C. T., Artificial neural networks: implications for the pharmaceutical sciences, *Drug Dev. Ind. Pharm.*, 21, 119, 1995.

7. Murtoniemi, E. et al., The advantages by the use of neural networks in modelling the fluidised bed granulation process, *Int. J. Pharm.*, 108, 15, 1994.

8. Merrku, P. and Yliruusi, J., Use of 3^3 factorial design and multilinear stepwise regression analysis in studying the fluidised bed granulation process, *Eur. J. Pharm. Biopharm.*, 39, 75, 1993.

9. Chen, Y. X., et al., Comparison of four artificial neural network software programs used to predict *in vitro* dissolution of controlled release tablets, *Pharm. Dev. Technol.*, 7, 373, 2002.

10. Plumb, A. P. et al., The effect of experimental design on the modelling of a tablet coating formulation using artificial neural networks, *Eur. J. Pharm. Sci.*, 16, 281, 2002.

11. Bourquin, J. et al., Pitfalls of artificial neural network (ANN) modelling technique for data sets containing outlier measurements using a study on mixture properties of a direct compressed dosage form, *Eur. J. Pharm. Sci.*, 7, 17, 1998.

12. Derringer, G. and Suich, R., Simultaneous optimisation of several response variables, *J. Qual. Tech.*, 12, 214, 1980.

13. Takayama, K., Fujikawa, M., and Nagai, T., Artificial neural network as a novel method to optimise pharmaceutical formulations, *Pharm. Res.*, 16, 1, 1999.

Appendix 1   Statistical Tables

Full sets of mathematical tables for techniques used in this book are widely available in reference sources and textbooks on statistics. Hence, the tables given here are those that are referred to in the worked examples in the chapters of this book.

A1.1   THE CUMULATIVE NORMAL DISTRIBUTION (GAUSSIAN DISTRIBUTION)

A normal distribution occurs with an infinite number of random events and can be represented by a plot of the magnitudes of the events (X axis) against their frequencies of occurrence (Y axis). The normal distribution is a theoretical concept, but it is followed in practice by large populations of random events. The plot is bell shaped, in which the maximum coincides with the arithmetic mean of the events, together with the median and the mode. The distribution is defined by (A1.1).

$$Y = \frac{1}{\sqrt{2\Pi}} \exp\left(-0.5z^2\right) \qquad (A1.1)$$

where
$z =$ the normal deviate, defined as the difference between the event size of interest and the universe mean, divided by the universe standard deviation.

Integration of (A1.1) gives the results shown in Table A1.1. These are the fractions of the total number of events represented by events within a particular size range. Thus, for example, in a normal distribution over 95% of the events lie between $\mu-1.96$ and $\mu+1.96$.

A1.2   STUDENT'S t DISTRIBUTION

Normal deviates can only be used when the universe mean and standard deviation are known. These values are not usually available, and the mean and standard deviation of the experimental sample must be used instead. These will be similar to the universe parameters, or even identical. Student's t values (Table A1.2) must be used under these circumstances instead of the normal deviates provided in Table A1.1.

TABLE A1.1
Values of the Normal Deviate

P'	z
0.05	0.06
0.10	0.13
0.20	0.25
0.30	0.39
0.40	0.52
0.50	0.67
0.60	0.84
0.70	1.04
0.80	1.28
0.90	1.65
0.95	1.96
0.99	2.58
0.995	2.81
0.999	3.29

TABLE A1.2
Values of Student's t

Degrees of Freedom (φ)	t Value			
	$P'=0.05$	$2P'=0.05$	$P'=0.01$	$2P'=0.01$
1	6.31	12.7	31.8	63.7
2	2.92	4.30	6.97	9.92
3	2.35	3.18	4.54	5.84
4	2.13	2.78	3.75	4.60
5	2.02	2.57	3.37	4.03
6	1.94	2.45	3.14	3.71
7	1.89	2.36	3.00	3.50
8	1.86	2.30	2.90	3.36
9	1.83	2.26	2.82	3.25
10	1.81	2.23	2.76	3.17
12	1.78	2.18	2.68	3.05
15	1.75	2.13	2.60	2.95
18	1.73	2.10	2.55	2.88
20	1.72	2.09	2.53	2.85
25	1.71	2.06	2.49	2.79
30	1.70	2.04	2.46	2.75
40	1.68	2.02	2.42	2.70
50	1.68	2.01	2.40	2.68

continued

TABLE A1.2 continued

Degrees of Freedom	t Value			
(φ)	$P'=0.0$ 5	$2P'=0.0$ 5	$P'=0.0$ 1	$2P'=0.0$ 1
60	1.67	2.00	2.39	2.66
70	1.67	1.99	2.38	2.65
80	1.66	1.99	2.37	2.64
90	1.66	1.99	2.37	2.63
100	1.66	1.98	2.36	2.63
120	1.66	1.98	2.36	2.62
∞	1.66	1.96	2.36	2.58

A1.3   ANALYSIS OF VARIANCE

The Student's t value is used to compare the means of two sets of data. When the means of more than two groups are to be compared, analysis of variance is employed. The statistical parameter F is calculated (Chapter 2) and compared with tabulated values of F. Critical values of F are also used in regression analysis (Chapter 4).

TABLE A1.3
Upper 5% Values of the *F* Distribution

Degrees of Freedom in Denominator	Degrees of Freedom in Numerator				
	1	2	3	4	5
1	161	200	216	225	230
2	18.5	19.0	19.2	19.2	19.3
3	10.1	9.55	9.28	9.12	9.01
4	7.71	6.94	6.59	6.39	6.26
5	6.61	5.79	5.41	5.19	5.05
6	5.99	5.14	4.76	4.53	4.39
8	5.32	4.46	4.07	3.84	3.69
10	4.96	4.10	3.71	3.48	3.33
15	4.54	3.68	3.29	3.06	2.90
20	4.35	3.49	3.10	2.87	2.71
27	4.21	3.35	2.96	2.73	2.57
30	4.17	3.32	2.92	2.69	2.53
40	4.08	3.23	2.84	2.61	2.45
45	4.06	3.21	2.82	2.59	2.43
50	4.03	3.18	2.79	2.56	2.40
100	3.94	3.09	2.70	2.46	2.31
∞	3.84	3.00	2.60	2.37	2.21

TABLE A1.4
Upper 1% Values of the *F* Distribution

Degrees of Freedom in Denominator	Degrees of Freedom in Numerator				
	1	2	3	4	5
1	4052	4999	5203	5625	5764
2	98.5	99.0	99.2	99.2	99.3
3	34.1	30.8	29.5	28.7	28.2
4	21.2	18.0	16.7	16.0	15.5
5	16.3	13.3	12.1	11.4	11.0
6	13.8	10.9	9.78	9.15	8.75
8	11.3	8.65	7.59	7.01	6.63
10	10.0	7.56	6.55	5.99	5.64
15	8.68	6.36	5.42	4.89	4.56
20	8.10	5.85	4.94	4.43	4.10
27	7.68	5.49	4.60	4.11	3.78
30	7.56	5.39	4.51	4.02	3.70
40	7.31	5.18	4.31	3.83	3.51
45	7.24	5.12	4.25	3.77	3.46
50	7.17	5.06	4.20	3.72	3.41
100	6.90	4.82	3.98	3.51	3.21
∞	3.94	3.09	2.70	2.46	2.31

Appendix 2　Matrices

Matrices have been used from time to time in this book. This section is written for the benefit of readers who are not familiar with the subject and is provided to make the book easier to understand.

A2.1 INTRODUCTION

Equations (A2.1) and (A2.2) are an example of a pair of simultaneous equations which can be solved to evaluate x and y. Solution of the equations first involves multiplication of each of the terms in (A2.1) by the coefficient of x in (A2.2), which is 2, followed by multiplication of each of the terms in (A2.2) by the coefficient of x in (A2.1), which is 4.

$$4x+y=8 \tag{A2.1}$$

$$2x+3y=12 \tag{A2.2}$$

This yields (A2.3) and (A2.4)

$$8x+2y=16 \tag{A2.3}$$

$$8x+12y=48 \tag{A2.4}$$

Subtraction of (A2.3) from (A2.4) then gives (A2.5)

$$10y=32 \tag{A2.5}$$

Therefore,

$$y=3.2$$

Substitution in (A2.1) or (A2.2) then gives $x=1.2$.

This elementary mathematical procedure is also the basis of the concept of matrices.

Equations (A2.1) and (A2.2) can be regarded in another way. For the coefficients 4, 1, 2, and 3, and for the solution $x=1.2$ and $y=3.2$, there are only two possible values on the right-hand sides of (A2.1) and (A2.2), namely, 8 and 12. These two values form the linear mapping of the left-hand sides of (A2.1) and (A2.2) when $x=1.2$ and $y=3.2$.

The study of matrices is less concerned with solutions than with relationships between coefficients, expressed for (A2.1) and (A2.2) in the form

$$\begin{bmatrix} 4 & 1 \\ 2 & 3 \end{bmatrix}$$

This block of numbers is an example of a matrix, generally defined as a rectangular array of numbers. Each number in the array is called an element, each set of elements running along a matrix is a row, and each vertical set of elements is a column. The above example is a 2×2 matrix, because it has two rows and two columns. It is also a square matrix, because the number of rows equals the number of columns. A matrix with n rows and n columns is called an nth-order matrix. Matrices are traditionally surrounded by square brackets, as shown above. In studies involving matrices, the elements form the data under investigation.

A single row of elements enclosed in square brackets, for example,

$$\begin{bmatrix} 1.0 & 1.4 & 1.2 & 1.5 & 1.3 \end{bmatrix}$$

is called a row vector, and a column enclosed in square brackets, for example,

$$\begin{bmatrix} 1.0 \\ 1.4 \\ 1.2 \\ 1.5 \\ 1.3 \end{bmatrix}$$

is a column vector.

Matrices can be of any size.

Equations (A2.1) and (A2.2) can be written in matrix form, as shown in (A2.6)

$$\begin{bmatrix} 8 \\ 12 \end{bmatrix} = \begin{bmatrix} 4 & 1 \\ 2 & 3 \end{bmatrix} \begin{bmatrix} 1.2 \\ 3.2 \end{bmatrix} \tag{A2.6}$$

or in general terms

$$\begin{bmatrix} x' \\ y' \end{bmatrix} = \begin{bmatrix} 4 & 1 \\ 2 & 3 \end{bmatrix} \begin{bmatrix} x \\ y \end{bmatrix} \tag{A2.7}$$

Equation (A2.7) tells us that if the column matrix

$$\begin{bmatrix} x \\ y \end{bmatrix}$$

is multiplied by the square matrix

$$\begin{bmatrix} 4 & 1 \\ 2 & 3 \end{bmatrix}$$

the answer will be the column matrix

$$\begin{bmatrix} x' \\ y' \end{bmatrix}$$

The matrix algebra procedure for multiplication of the square matrix by the column matrix can be derived by comparison with the classical method of solving simultaneous equations. Thus, substitution for $x = 1.2$ and $y = 3.2$ in (A2.1) and (A2.2) gives

$$x' = (4 \times 1.2) + (1 \times 3.2) = 8 \tag{A2.8}$$

and

$$y' = (2 \times 1.2) + (3 \times 3.2) = 12 \tag{A2.9}$$

Thus, to multiply the column vector and the 2×2 matrix in (A2.6),

1. x' is equal to the product of the first element in the first row of the 2×2 matrix and the top element of the column vector on the right-hand side, giving 4×1.2, plus the product of the second element of the top row of the 2×2 matrix and the bottom element of the column vector on the right-hand side, to give 1×3.2, yielding a total of $(4 \times 1.2) + (1 \times 3.2) = 8$.
2. The y coordinate is the product of the first element of the bottom row of the 2×2 matrix and the top element of the column vector on the right-hand side, plus the product of the second element of the bottom row of the 2×2 matrix and the bottom element of the column vector on the right-hand side, yielding a total of $(2 \times 1.2) + (3 \times 3.2) = 12$.

Matrices can be subjected to other mathematical manipulations, such as addition and subtraction, but as with multiplication, the procedures involve different rules from classical algebra. Some of these are outlined below.

A2.2 ADDITION AND SUBTRACTION OF MATRICES

These processes can only be carried out between matrices having the same order. The procedures follow logically from classical mathematics, in that each element in the right-hand matrix is added to or subtracted from its corresponding element in the left-hand matrix. Thus, for example, to add the following 2×2 matrices,

$$\begin{bmatrix} a_{11} & a_{12} \\ a_{21} & a_{22} \end{bmatrix} + \begin{bmatrix} b_{11} & b_{12} \\ b_{21} & b_{22} \end{bmatrix} = \begin{bmatrix} (a_{11}+b_{11}) & (a_{12}+b_{12}) \\ (a_{21}+b_{21}) & (a_{22}+b_{22}) \end{bmatrix}$$ (A2.10)

and to subtract

$$\begin{bmatrix} a_{11} & a_{12} \\ a_{21} & a_{22} \end{bmatrix} - \begin{bmatrix} b_{11} & b_{12} \\ b_{21} & b_{22} \end{bmatrix} = \begin{bmatrix} (a_{11}-b_{11}) & (a_{12}-b_{12}) \\ (a_{21}-b_{21}) & (a_{22}-b_{22}) \end{bmatrix}$$ (A2.11)

The same procedures apply to larger matrices, for example,

$$\begin{bmatrix} 4 & 2 & 1 \\ 3 & 4 & 2 \\ 6 & 3 & 6 \end{bmatrix} + \begin{bmatrix} 1 & 2 & 3 \\ 4 & 5 & 6 \\ 7 & 8 & 9 \end{bmatrix} = \begin{bmatrix} 5 & 4 & 4 \\ 7 & 9 & 8 \\ 13 & 11 & 15 \end{bmatrix}$$

A2.3 MULTIPLICATION OF MATRICES

Multiplication procedure varies with the functions that are being multiplied.

A2.3.1 MULTIPLYING A MATRIX BY A CONSTANT

This process is represented by placing the constant (b) outside the brackets, as shown below, and follows the logical course of multiplying all the elements by the constant, as shown in (A2.12)

$$b \begin{bmatrix} a_{11} & a_{12} \\ a_{21} & a_{22} \end{bmatrix} = \begin{bmatrix} ba_{11} & ba_{12} \\ ba_{21} & ba_{22} \end{bmatrix}$$ (A2.12)

The same procedure applies to larger matrices, for example,

$$5 \begin{bmatrix} 4 & 2 & 1 \\ 3 & 4 & 2 \\ 6 & 3 & 6 \end{bmatrix} = \begin{bmatrix} 20 & 10 & 5 \\ 15 & 20 & 10 \\ 30 & 15 & 30 \end{bmatrix}$$

A2.3.2 MULTIPLICATION OF ONE MATRIX BY ANOTHER

Multiplication of matrices is related to the solving of simultaneous equations and hence to the derivation of regression equations used in Chapters 4, 7, and 8.

Consider the two matrices A and B. Matrix A can only be multiplied by matrix B if the number of columns in matrix A is the same as the number of rows in matrix B. The resultant matrix (C) will have the same number of rows as matrix A and the same number of columns as matrix B.

An element c_{ij} in matrix C is obtained from row i of matrix A and column j of matrix B by multiplying the corresponding elements together, followed by addition of all the products. Thus, in general terms, each element of the product matrix is calculated according to (A2.13)

$$c_{ij}=a_{i1}\times b_{1j}+a_{12}\times b_{2j}+\cdots+a_{ik}\times b_{kj} \tag{A2.13}$$

The multiplication of two 2×2 matrices is shown in (A2.14). The result is another 2×2 matrix.

$$\begin{bmatrix}4 & 2\\3 & 1\end{bmatrix}\times\begin{bmatrix}5 & 7\\6 & 8\end{bmatrix}=\begin{bmatrix}[(4\times5)+(2\times6)] & [(4\times7)+(2\times8)]\\ [(3\times5)+(1\times6)] & [(3\times7)+(1\times8)]\end{bmatrix}=\begin{bmatrix}32 & 44\\21 & 29\end{bmatrix} \tag{A2.14}$$

The product of a 4×4 matrix and a 4×2 matrix is a matrix with four rows and two columns (A2.15), each element of which is calculated from (A2.13)

$$\begin{bmatrix}a_{11} & a_{12} & a_{13} & a_{14}\\a_{21} & a_{22} & a_{23} & a_{24}\\a_{31} & a_{32} & a_{33} & a_{34}\\a_{41} & a_{42} & a_{43} & a_{44}\end{bmatrix}\times\begin{bmatrix}b_{11} & b_{12}\\b_{21} & b_{22}\\b_{31} & b_{32}\\b_{41} & b_{42}\end{bmatrix}=\begin{bmatrix}c_{11} & c_{12}\\c_{21} & c_{22}\\c_{31} & c_{32}\\c_{41} & c_{42}\end{bmatrix} \tag{A2.15}$$

Equation (A2.15) shows the multiplication of a 4×4 matrix (matrix A) and a 4×2 matrix (matrix B). This is possible because the number of columns in matrix A equals the number of rows in matrix B. It is not possible to multiply matrix B by matrix A, because the number of columns in matrix B does not equal the number of rows in matrix A.

Furthermore, in classical mathematics, a product of two numbers is the same, irrespective of the order in which the numbers are taken. For example, $a\times b$ is equal to $b\times a$. This does not always apply in matrix algebra. Thus, reversing the order of the matrices on the left-hand side of (A2.14) gives a different matrix from that shown, as demonstrated below, using simple numbers

$$\begin{bmatrix}4 & 2\\3 & 1\end{bmatrix}\times\begin{bmatrix}5 & 7\\6 & 8\end{bmatrix}=\begin{bmatrix}32 & 44\\21 & 29\end{bmatrix}$$

but

$$\begin{bmatrix}5 & 7\\6 & 8\end{bmatrix}\times\begin{bmatrix}4 & 2\\3 & 1\end{bmatrix}=\begin{bmatrix}41 & 17\\48 & 20\end{bmatrix}$$

A2.3.3 MULTIPLICATION BY A UNIT MATRIX

In any square matrix, the elements running diagonally from the top left-hand corner to the bottom right-hand corner form the leading diagonal. Multiplication by a matrix in which all the elements in the leading diagonal are the same and the remaining elements are all zero is equivalent to multiplying by the diagonal element alone, as, for example, in (A2.16)

$$\begin{bmatrix} a_{11} & a_{12} \\ a_{21} & a_{22} \end{bmatrix} \times \begin{bmatrix} b & 0 \\ 0 & b \end{bmatrix} = \begin{bmatrix} (ba_{11}+0) & (0+ba_{12}) \\ (ba_{21}+0) & (0+ba_{22}) \end{bmatrix} = \begin{bmatrix} ba_{11} & ba_{12} \\ ba_{21} & ba_{22} \end{bmatrix} \qquad \text{(A2.16)}$$

which is the solution given by (A2.12).

A square matrix in which all the elements in the leading diagonal are equal to 1 and the remainder is equal to zero is called a unit matrix. The unit matrix is the matrix equivalent to unity in classical mathematics, because if a matrix is multiplied by a unit matrix, the answer will be the original matrix, as shown in (A2.17)

$$\begin{bmatrix} 4 & 1 \\ 2 & 3 \end{bmatrix} \times \begin{bmatrix} 1 & 0 \\ 0 & 1 \end{bmatrix} = \begin{bmatrix} [(4\times1)+(1\times0)] & [(4\times0)+(1\times1)] \\ [(2\times1)+(3\times0)] & [(2\times0)+(3\times1)] \end{bmatrix} = \begin{bmatrix} 4 & 1 \\ 2 & 3 \end{bmatrix} \qquad \text{(A2.17)}$$

In general,

$$\begin{bmatrix} X & 0 \\ 0 & X \end{bmatrix}$$

where X is a constant, is equal to X.

A2.3.4 MULTIPLICATION BY A NULL MATRIX

All the elements of a null matrix are zero. For example,

$$\begin{bmatrix} 0 & 0 & 0 \\ 0 & 0 & 0 \\ 0 & 0 & 0 \end{bmatrix}$$

is a third-order null matrix. The null matrix is the matrix equivalent of zero in classical mathematics. The product of any matrix with the null matrix of the same order is equal to zero.

A2.3.5 TRANSPOSITION OF MATRICES

Each element in a matrix is defined by its row and column number; for example, in the matrix C, element C_{ij} is in row i and column j. The transpose matrix is obtained by exchanging rows and columns, so that the element C_{ij} becomes element C'_{ji} in the transpose matrix C'.

Thus, the transposition of the matrix

$$\begin{bmatrix} 1 & 2 & 3 \\ 4 & 5 & 6 \end{bmatrix}$$

is

$$\begin{bmatrix} 1 & 4 \\ 2 & 5 \\ 3 & 6 \end{bmatrix}$$

These matrices satisfy the conditions whereby they can be multiplied together. The product $C'C$ is called the information matrix and in the above example has the form

$$\begin{bmatrix} 14 & 32 \\ 32 & 77 \end{bmatrix}$$

It is a square matrix and is symmetrical, in that the top right element is equal to the bottom left element.

A2.3.6 INVERSION OF MATRICES

If the product of matrix A and matrix B is a unit matrix, then matrix B is the inverse of matrix A. Thus, if $A \times B = I$, then $B = A^{-1}$. For inversion of matrix A to be possible, then A must be a square matrix and its determinant must not be equal to zero (see below).

The transpose of the inverse of matrix A is identical to the inverse of its transpose matrix, that is,

$$(A')^{-1} = (A^{-1})'$$

A2.4 DETERMINANTS

Matrix algebra is a relatively recent science in terms of the history of mathematics. The word matrix was first used in mathematics in 1850 by Sylvester, whereas the use of determinants is older, originating with Leibniz in 1693. Determinants are expressions associated with square arrays of numbers and were originally used to solve simultaneous linear equations. The traditional way of solving such equations has been demonstrated with (A2.1) and (A2.2) above.

The procedure involving the terms on the left-hand sides was

$$(4 \times 2) + (4 \times 3) - (2 \times 4) - (2 \times 1)$$

which reduces to

$$(4 \times 3) - (2 \times 1) = 10$$

The procedure can be expressed in the form

$$\begin{vmatrix} 4 & 1 \\ 2 & 3 \end{vmatrix}$$

The vertical lines on each side of the numbers indicate that the expression is the determinant of the corresponding matrix.

The determinant is solved by subtracting the product of the second element in the first row and the first element in the second row from the product of the first element in the first row and the second element in the second row. It is a second-order determinant, because it has two rows and two columns and, in this case, solves to 10.

In general terms, the determinant of the matrix

$$\begin{bmatrix} a_{11} & a_{12} \\ a_{21} & a_{22} \end{bmatrix} = \left(a_{11} \times a_{22} \right) - \left(a_{21} \times a_{12} \right)$$

An important characteristic of determinants is that when the elements in two or more columns of a matrix are related in the same way, its determinant reduces to zero. Thus, in the matrix

$$\begin{vmatrix} 2 & 3 \\ 4 & 6 \end{vmatrix}$$

both elements in row 2 are twice the value of the elements above them, and because of this proportionality, the determinant is zero $[(2 \times 6) - (4 \times 3) = 0]$. This property is used in multivariate analysis as a test for relationships between columns of elements.

Determinants of 3×3 matrices are more difficult to calculate. Each element is multiplied in turn by the determinant of the 2×2 matrix whose elements are neither in the same row nor in the same column as the first row element.

The result for the second element in the top row is then subtracted from the sum of the other two results. Thus, taking the following matrix as an example:

$$\begin{bmatrix} a_{11} & a_{12} & a_{13} \\ a_{21} & a_{22} & a_{23} \\ a_{31} & a_{32} & a_{33} \end{bmatrix}$$

$$\text{determinant} = a_{11}[(a_{22} \times a_{33}) - (a_{23} \times a_{32})] - a_{12}[(a_{21} \times a_{33}) - (a_{23} \times a_{31})] + a_{13}[(a_{21} \times a_{32}) - (a_{22} \times a_{31})]$$

The directions of the signs between the second-order determinants follow logically from the classical method of solving simultaneous equations. The order in which the elements are taken is also important. The columns must be represented in the lower-order determinants in the same way as they appear in the original matrix.

The value of a third-order determinant in establishing relationships between variables can be illustrated by using the results in Table A2.1. This gives the diffusion coefficients of 4-hydroxybenzoic acid in three gelatin gels, A, B, and C, together with the microscopic and macroscopic viscosities of the gels.[1] The objective of the experiment was to ascertain whether macroviscosity or microviscosity influenced diffusion. Observation of the results is all that is needed to give the answer to

TABLE A2.1
The Influence of Viscosity on the Migration of 4-Hydroxybenzoic Acid through Glycerogelatin Gels[1]

Sample	Diffusion Coefficient $(mm^2 \cdot h^{-1})$	Microscopic Viscosity $(Ns \cdot m^{-2} \times 10^3)$	Macroscopic Viscosity $(Ns \cdot m^{-2} \times 10^3)$
A	0.021	13.30	2.20
B	0.040	6.52	20.2
C	0.027	10.86	26.8
Mean	0.0293	10.23	16.40
Standard deviation	0.0097	3.434	12.73

this question, so that statistical methods need not be used to establish relationships in this case. However, by looking at such a simple situation, it can be seen how the methodology can be applied to more complicated problems. A matrix of the standardized values of data from Table A2.1 is shown in Table A2.2.

The calculation of the determinant of this matrix is shown in (A2.18).

$$\begin{vmatrix} -0.8557 & 0.8940 & -1.1155 \\ 1.1031 & -1.0804 & 0.2985 \\ -0.2371 & 0.1835 & 0.8170 \end{vmatrix}$$

$$= -0.8557 \begin{vmatrix} -1.0804 & 0.2985 \\ 0.1835 & 0.8170 \end{vmatrix} - 0.8940 \begin{vmatrix} 1.1031 & 0.2985 \\ -0.2371 & 0.8170 \end{vmatrix} + (-1.1155) \begin{vmatrix} 1.1031 & -1.0804 \\ -0.2371 & 0.1835 \end{vmatrix} \quad (A2.18)$$

$$= [0.8557 \times (-0.9375)] - (-0.8940 \times 0.9720) + [-1.1155 \times (-0.0537)]$$

$$= 0.8022 - 0.8690 + 0.0599$$

$$= -0.0069$$

The determinant is very small (−0.0069), signifying that at least two of the columns are related. The number and nature of the columns which are related can be assessed by calculating the determinants of the second-order matrices. There are three second-order determinants involving the diffusion coefficient and microviscosity. The largest of these, ignoring the sign, is 0.0617. This indicates that the diffusion coefficient and the microviscosity are related in some way, that is,

TABLE A2.2
Standardized Values of the Data in Table A2.1

Sample	Diffusion Coefficient	Microscopic Viscosity	Macroscopic Viscosity
A	−0.8557	0.8940	−1.1155
B	1.1031	−1.0804	0.2985
C	−0.2371	0.1835	0.8170

the microviscosity is a predictor of the diffusion coefficient. Direct proportionality would give a value of zero for all three determinants, but this is not achieved because of experimental scatter.

There are also three second-order determinants involving the diffusion coefficient and macroviscosity. The smallest of these, ignoring the sign, is 0.964, indicating that the diffusion coefficient is not related to macroviscosity. Also, all the three second-order determinants of microviscosity and macroviscosity have values in excess of 0.9, indicating that there is no relation between these either.

The determinant of a fourth-order matrix is obtained by multiplying each element in the first row by the determinant of the 3×3 matrix with which it shares neither a row nor a column. The results for the first and third elements in the first row are added together, and the second and fourth row element results subtracted from the total.

Determinants are tedious to calculate, but many spreadsheet packages contain facilities to calculate determinants and manipulate data in matrix form. For example, in Microsoft Excel®, the command MDETERM returns the matrix determinant of an array of numbers and MMULT returns the matrix product of two arrays.

FURTHER READING

Bronson, R., *Matrix Methods, An Introduction*, 2nd ed., Academic Press, New York, 1991.
Coulson, A. E., *Introduction to Matrices*, Longman, London, 1965.

REFERENCE

1. Armstrong, N. A. et al., The influence of viscosity on the migration of chloramphenicol and 4-hydroxybenzoic acid through glycerogelatin gels, *J. Pharm. Pharmacol.*, 39, 583, 1987.

Index

Addition of matrices, 225–6
Alphabetical notation of experimental
    designs, 91, 104
Analysis
    cluster, 63–5
    discrimination, 67–70
    factor, 75–9
    least squares, 35
    multiple regression, 44–7
    multivariate, 55, 60, 63, 75
    principal components, 70–5
    regression, 5, 40, 41, 176, 185
    sequential, 173–7
Analysis of variance
    one-way, 19–21
    two-way, 22–4
    with factorial design, 94–100
ANN, *see* artificial neural networks
ANOVA, *see* analysis of variance
Artificial neural networks
    architecture, 212
    hidden layer, 210, 214
    input layer, 210
    output layer, 210
    pharmaceutical applications of, 212–5
    training, 210, 211
Asymmetric designs, 114–15

Backpropagation, 211
Barriers in sequential analysis, 175–6, 177
Blocked designs
    for four factor, two level experiments, 117,
        118, 120
    for three factor, three level experiments,
        118
    for three factor, two level experiments, 117
    for two factor, two level experiments, 116
Boundaries
    in experimental designs, 136–7
    in sequential analysis, 175–7

Box-Behnken designs
    for a four factor experiment, 126
    for a three factor experiment, 126, 127
Box plot, 16, 17
Brainmaker©, 214

CAD/CHEM©, 214
Cartesian plots, 63–5
Central composite designs
    for a three factor experiment, 124, 125
    for a two factor experiment, 122–3
Centre of gravity design, 125
CHEOPS®, 4
Christmas tree boundary, 175
Cluster
    analysis, 63–6
    plot
        one dimensional, 65
        two dimensional, 65
Coded data, 93, 123
CODEX®, 4
Coefficient
    correlation, 47–8
    of determination, 38–40
    of multiple regression, 47–8
    of rank order correlation, 79
    standard error of, 40, 48
Column vector, 224, 225
Communality, 73, 77
Comparison of mean values
    among more than two groups of data,
        18–19
    when the variance of the whole population
        is known, 10–2
    when the variance of the whole population
        is not known, 12–5
Comparison of model-independent and
        model-dependent methods, 184–7
Computers and experimental design, 2–4
Confidence interval, 11–12, 16

234 Index

Confounding, 116–17, 120, 121, 153
Connectionism, 209
Constraint, 136–7
Contour plots
  and optimization, 163–5
  for mixtures, 203, 204
Contraction, 181
Correction term, 19, 106,
  111–14
Correlation
  and causality, 52
  coefficient, 39, 47, 52
  matrix, 63, 72, 76, 79
  rank order, 79
Covariance
  matrix, 59–62
Critical values of F, 40, 221
Cubic curve
  reduced, 199, 201, 203
Cumulative normal distribution, 219
Curve fitting of non-linear relationships,
  41–4
Curve
  cubic, 43, 199, 201, 203
  exponential, 44
  geometric, 44, 167, 178
  parabolic, 42–3
  polynomial, 42, 44
  quadratic, 42–3, 147
Curved response surface, 143

D-efficiency, 129
Degrees of freedom, 37
Dendrograms, 65–6
Design-Ease®, 3
Design-Expert®, 3
Design space
  extension of, 161–3
Design validation, 1, 136, 139–40, 150
Desirability function
  overall, 167–8
  partial, 166–8
Determinants
  of a (2x2) matrix, 230, 231
  of a (3x3) matrix, 230, 232
  of a (4x4) matrix, 232
Determination, coefficient of,
  38–40
Discrimination analysis, 67–70
Dispersion matrix, 129

Distance
  Euclidean, 56, 70
  matrix, 55–9
  multivariate, 55–9
  root mean square, 69
Distance matrix, 55–9
Doehlert design
  for three factors, 128, 129
  for two factors, 127, 128
D-optimal design, 129

ECHIP®, 4
Efficiency of experimental designs, 129–30
Eigenvalues
  calculation of, 72–3, 74, 78
Eigenvectors
  calculation of, 72–3, 74, 78
Element, 57, 60–1, 72–3, 225–6, 227–9
Equation
  cubic, 43
  parabolic, 42–3
  polynomial, 42
  quadratic, 42–3, 147, 200
  reduced cubic, 199, 201
Error
  of the coefficient, standard, 40, 48
  of the intercept, standard, 48
Euclidean distance, 70
Excel®, 3, 37, 232
Expansion, 179, 181
Experimental designs for mixtures, 189–208
Experimental domain, 136–7
Experimental units, 89, 90, 137, 140,
  154, 158
Expert systems, 209
Exponential curve, 44
Extension of the design space, 161–3
Extrapolation, 36, 150, 212

F value
  tables, 37, 40, 60, 108
Factor analysis, 75–9
Factor interaction, 87, 88, 90, 91, 93,
  94, 95, 99
Factorial design of experiments
  and analysis of variance, 94–100
  and linear regression, 98–100
  and optimization, 89, 122
  and Yates's treatment, 95–8
  asymmetric designs, 114

blocked factorial designs, 115–18
Box-Behnken designs, 126–7
central composite designs, 122–5
center of gravity designs, 125
Doehlert designs, 127–9
efficiency of, 129–30
experimental sequence, 103–4
five factor, two level fractional designs, 154
five factor, two level full designs, 153
four factor, two level fractional designs, 117, 118, 120
four factor, two level full designs, 152
fractional designs, 119–20
interaction between factors, 86–9
linear regression, 98–100
mixed designs, 114–15
more than three factors, 150–4
notation, 89–91
Plackett-Burman designs, 121–2
replicated designs, 100–3
screening designs, 119, 122, 136
standard order, 91, 92, 95, 103, 104, 108
three factor, three level designs, 110–15
three factor, two level designs, 92, 93–4, 96, 97, 101–2, 104, 117
three level designs, 108
two factor, three level designs, 105, 106, 107
two factor, two level designs, 84–9
FFDESIGN, 3
FFACTORIAL, 3
Fixed sample tests, 173, 177
Four component diagrams, 193, 194
Four factor Box-Behnken designs, 126
Freedom, degrees of, 37–8
Furthest neighbour plot, 66

Gaussian distribution, 219
Geometric curve, 178
Goal seeking, 173

Half design, 119, 153
Hampel's Rule, 17

Information matrix, 129, 229
Interaction between factors in factorial designs
graphical detection, 86
quantitative estimation, 87

Interaction between factors in optimization, 89, 122
Interaction between independent variables, 48–9
Intercept
standard error of, 48
Inversion of matrices, 229

Kolmogorov theorem, 212, 213, 215

Leading diagonal, 59, 63, 72, 77, 227
Least significant difference, 21
Least squares analysis, 35
Linear regression
and factorial design, 98–100
multiple, 98
Logarithmic relationship, 44, 60
Lotus 1-2-3®, 3
Lower boundary in sequential analysis, 175, 177

Mann-Whitney U-test, 25
Mathematical modeling, 158
Matrices
addition of, 225–6
determinants of, 229–32
multiplication of, 226–8
subtraction of, 225–6
transposition of, 228–9
Matrix
correlation, 62–3
covariance, 59–62
distance, 55–9
moment, 129–30
square, 62, 76, 227, 228
MDETERM, 232
Mean squared distance, 69
Mean values, comparison of, 9–24
MINITAB®, 3
Mixed designs, 115
Mixtures
and contour plots, 197–8, 200–1, 203, 204
and Pareto-optimality, 203–5
and process variables, 205–7
experimental designs for, 189–207
optimization of, 202–3
pseudocomponents in, 192–3, 195
with more than three components, 193–5
with three components, 190–3
with two components, 190–1

MMULT, 232
Model-dependent optimization
    comparison with model-independent
        optimization, 184–7
    for mixtures, 195
Model equation, 143, 198, 203, 215
Model-independent optimization, 177–84
    comparison with model-dependent
        optimization, 184–7
    simplex search, 177–84
Moment matrix, 129–30
Multicriteria decision making, 157
Multiple regression analysis, 44–8
Multiple regression, coefficient of, 47–8
Multiplication of matrices
    by a constant, 226
    by a null matrix, 228
    by a unit matrix, 227–8
    by another matrix, 226–7
Multivariate distances, 55–9
Multivariate analysis, 55, 60, 63, 75

Neighbours
    furthest, 66
    nearest, 65–6
Nearest neighbour plot, 66
Neural computing, 209, 212
NeuralWorks Professional II/Plus©, 214
Neuroshell 2©, 214
Nonparametric methods
    for paired data
        sign test, 25–7
        Wilcoxon signed rank test, 27–8
    for unpaired data
        Wilcoxon two-sample test, 29–32
Normal deviate (z), 219
Normalization, 183, 184
Normalization in model-independent
    optimization, 183–5
Notation in experimental design
    alphabetical, 91
    numerical, 104
Numerical notation in experimental design,
    104

One-way analysis of variance, 19
Optimization
    and response surface methodology, 135
    by combining contour plots, 163–5
    by Pareto-optimality, 169–71
    by simplex search, 177–84
    constraints, 158, 161, 163, 164
    extrapolation, 161, 163, 185
    model-dependent, 157–71
    model-independent, 173–87
    of mixtures, 202–3
    use of coded data, 90, 123
    with interaction between factors, 86–9
Outlying data points, 15–17
Outlying results, 15–16
Overall desirability function, 167

Parabola, 42–3
Parallel distributed processing, 209
Pareto-optimal plots, 169
Pareto-optimal points, 170, 204, 206
Pareto-optimality
    with mixtures, 203
Partial desirability function, 167–8, 215
Path of steepest ascent, 162–3
PBDESIGN, 3
Pharmaceutical applications of artificial
    neural networks, 212–15
Plackett-Burman designs, 121–2
Planar response surface, 146
Polynomial curve, 42, 44
Pooled variance, 9, 13
Power series, 42
Principal component, 70, 72, 74
Principal components analysis, 70–5
Process variables in mixture experiments,
    205
Pseudocomponents in mixture experiments,
    192–3, 195

Quadratic equation, 43, 147, 161, 200
Quadratic relationship, 42–3

Rank order correlation, 79
Rank order correlation coefficient, 79
Reduced cubic equation, 199, 201
R-efficiency, 129
Reflection, 181
Regression
    analysis, 5, 42, 44–6, 98, 158
    coefficient, 98, 99, 100
    equation, 36–7, 100, 145, 197, 213
    line, 35–7, 41
    least squares, 33, 34, 35
    linear, 33–7

multiple
   coefficient of, 47–8
   nonlinear, 49
   stepwise, 49–50
   sum of squares, 37, 38, 40, 47, 49
Regression analysis, 5, 42, 44–6,
   98, 158
Replicated factorial designs, 100–3
Replicate determinations, 14, 27, 32
Residual sum of squares, 19, 23,
   40, 106
Residuals, 23–4, 35, 147
Response surface
   curved, 143
   planar, 143
Response surface methodology
   for five factors, 153
   for four factors, 152
   for three factors, 150–1
   from first order models, 137–43
   from higher order models, 203
   in experiments with mixtures, 195–201
Row vector, 224
RS/Discover®, 3

SAS®, 3
Saturated designs, 100, 150, 198, 199
Scatter diagram, 67–8
Screening designs, 119, 122, 136
Second order determinant, 230–2
Sequence of experiments, 103–4
Sequential analysis
   barriers, 175–6
   boundaries, 175–6, 177
   Wald diagram, 173–7
Sequential methods, 173–87
Sigmoid transfer function, 211
Sign test, 25–7
Simplex, 177–84
Simultaneous methods, 158
Six factor, two level fractional factorial
   designs, 120
Spearman coefficient of rank order
   correlation, 50
Spreadsheet packages, 232
SPSS®, 3
Square matrix, 228
Standard
   error of the coefficient, 48
   error of the intercept, 48

Standard order in experimental design, 92,
   95, 96, 103, 104
Standardization of data, 58, 59, 67, 71,
   72, 76, 231
Star design, 123
Star points, 123–5, 146, 153
Statgraphics Plus®, 4
Statistical tables, 219–22
Steepest ascent, path of, 162–3
Stepwise regression, 49–50
Student's t, 219–2
Student's t distribution, 219–2
Subtraction of matrices, 225–6

t test, *see* Student's t
Ternary diagrams, 190–3
Third order determinant, 230
Three component systems, 190–3
Three-dimensional diagram, 6, 44, 135
Three factor
   Box-Behnken designs, 126
   central composite designs, 123–5
   Doehlert designs, 128, 129
   factorial designs, 91–4
   three level factorial designs, 106, 107,
      110, 118
   two level factorial designs, 84, 86
Time trend, 104
Total sum of squares, 19, 22, 40
Transformations, 44
Transpose matrix, 228
Transposition of matrices, 228
Trimmed mean, 16
Truncation in sequential analysis,
   175, 177
Two factor
   central composite designs, 122–3, 124
   three level full factorial designs,
      105–7
   two level factorial designs, 84–6
   two level factorial designs with interaction,
      86–9
Two ingredient mixtures, 182,
   189–90
Two-way analysis of variance, 22, 106

U-test, 25
Univariate methods, 5, 55
Upper boundary in sequential analysis,
   173–5

Validation of experimental designs, 140
Variance, 10, 12, 19, 20, 22, 23, 40, 94, 97,
    102, 107, 109, 221
Variance ratio, 40
Vector
  column, 224, 225
  row, 224
Virial
  coefficients, 42
  equation, 42

Wald diagram, 173, 174, 175, 176, 177
Weighting factor, 183, 184, 185,
    186, 187
Wilcoxon signed rank test, 25, 27, 29
Wilcoxon two-sample test, 25, 29

Yates's treatment
  in replicated designs, 96, 97,
    100–3
  in three level designs, 108, 114, 118

Milton Keynes UK
Ingram Content Group UK Ltd.
UKHW040105071024
449327UK00019B/836

9 780367 391188